In My Life

Saundra Mathis-Copeland
Author

Outskirts Press, Inc.
http://www.outskirtspress.com

ISBN: 978-1-4327-9634-1

Outskirts Press and the "OP" logo are trademarks belonging to Outskirts Press, Inc.

PRINTED IN THE UNITED STATES OF AMERICA

Table of Contents

FOREWORD

For years, Sandy contemplated writing a book on the highs and lows of a woman whose life sometimes seemed to be spiraling out of her control. Her many life experiences have brought her to this moment in time when she is able to reveal, candidly, those experiences. Years ago, she was frightened of opening her life to public scrutiny. Now, she is willing—and this is the result.

In these pages, Sandy chronicles some of the fears, triumphs, defeats, and joys that have marked her life. Her goal is to inform and inspire other women to accept the cards that life has dealt them, with the sure knowledge that God is in control. This knowledge has helped her to heal, grow, endure, and love, in spite of difficult times and tough situations. She has come to understand that all of these life lessons were necessary to shape her into the woman that she is today.

This book contains a liberating message for all who read it, and it will empower them to embrace the vicissitudes of their own lives. It is a heartfelt, inspirational journey that is certain to make the reader want to hear more from this courageous, giving, and gifted woman.

Bishop Bernard N. Bragg

1. Pregnant With My First Blessed Child, Anisia

I never thought I would get pregnant before marriage. I grew up in a Pentecostal Apostolic church, and I was always taught to wait till you're married, In church, I've always been a leader or role model for the young people to look up to. I was a lead singer in the church choir and groups; I've always been the one to give good examples and learn to think for myself, and not to rely on other people's opinion all the time. I never thought that I was perfect, but we all fall short at some point in our lives—we are human. Some people used to put me on a pedestal, as if I could do no wrong. I've always been around friends who had children before they were married, but I've never judged them—who am I to judge? I always told myself that I would wait until I was totally committed, and that meant marriage first.

I started dating very young. Since the age of fourteen, it seemed that whenever I started dating, the boyfriends would come in groups of two—I don't mean to brag, but it was never one at a time. I love to sing—that's what I do best, and it is my gift from God. I have many talents, but singing is my first passion, and I believe that's what made me attractive to a lot of young men. Whenever you're in some kind of limelight, big or small, or your name is ringing out there, people seem to gravitate to you more. That was what happened to me in relationships at a young age. I was never afraid to belt out my loud singing voice, and never afraid to sing or perform in front of anybody—whenever I was in a relationship, the guy felt like he was dating a star! That was special at that time, but as I got older, I wanted to feel that guys were not dating me just for my voice, but for who I was as a person, inside and out.

I was always very open and very honest and truthful about myself. It was 1985, and I worked at the First National Bank of Boston (now known as Bank of America) on Morrissey Boulevard in Dorchester, Massachusetts. I worked in the Dividend Reinvestment shareholder department. I used to handle stocks and bonds. A particular young man always used to come by my department and just start smiling and staring at me for a couple of seconds.

Whenever I went on my lunch break, I would go to the cafeteria for lunch instead of going outdoors unless it was very nice out; they had very good cooks, therefore I didn't have a problem eating in the cafeteria. He used to go to lunch at the same time I did. I would always sit alone, but one day he found his way over and introduced himself. He said, "Hi, my name is Jamie, and I've been eyeing you for a while." We exchanged numbers. I was blown away by his personality. He had never heard me sing, nor did I tell him that I sang—I left that part out for a while.

As time went on, we grew closer to each other, and I started bringing him around my family. I thought it would make a big difference if he met my parents—of course, there was another young man who was after me for a while; his name was Stan, and he was my youngest brother Rodney's best friend. He would always say to my mother, "One day I'm going to marry your daughter when I get older." I would always laugh. Stan would always ask me, "Who are you talking to? I'm going to take you away from him." I thought that he was talking crazy; he'd been chasing me forever and a day, since back in high school, but I was head over heels about Jamie. He even gave me a nickname—"muffin"—that I thought was special between us.

As time went on, Stan realized I was getting serious with Jamie, and found out some information that he thought I should know…I don't know how he found out, but Stan came by the house late one evening to see me. I had just walked in the house, and he approached me. Every time Stan started a sentence, he would say something like "I need to talk to you." This time, he said, "Sweetheart, it's important. I already spoke with your brother, and he advised me to talk with you first." So he asked if we could take a drive and talk. At first I didn't want to hear what he had to say, but I looked at him in his green eyes—sometimes they would turn hazel, and I knew it was something serious he wanted to talk about, even though I knew he wanted me for himself.

So we took a spin in my car. When we started driving, I said, "Stan, what is it? Spit it out."

"Sweetheart, you're not going to believe this, but I have to protect you from the guy you are dating."

At first I thought he was jealous, but he looked at me without smiling, as if he had tears in his eyes—they were glassy, so I couldn't tell if he was for real. I asked him again, "What you are trying tell me?"

He said, "The guy you are dating is very dangerous, and I know him very well. He sliced his wife's face. Did you know he's married?"

"No way! But I knew he had a son."

"He's not good for you—at some point he will show his true colors. He's abusive to women, and I don't want you to get physically hurt, or me and your brother will do him in. You know how protective your brother is about his sisters."

Right then and there, I knew Stan was not playing around; he was very serious, and I believed him. I was in shock, because Jamie never hurt me nor disrespected me in any kind of way. That was a hard pill to swallow—believe it or not, Jamie spoiled me to death; he showered me with dinner and gifts all the time and always had kind words to say to me. I thought about all of that, still in shock.

Stan said, "Don't waste another breath on him anymore, or I'm stepping in." As we kept driving and talking, we saw Jamie walking; I was not sure where he was going, and I didn't care. I asked Stan to pull this freakin' car over—at the time I was so mad, it was another word I used!

Stan yelled, "He's a liar—leave him alone for good."

I said, "Pull over so I can go after him for not telling me he's married, and for being abusive to women." I yelled at Stan, "Turn the car around!"

He said, "Let me handle this—you stay in the car." Stan stopped the car, and kept his cool all along, I believe because of me. He didn't want the both of us going off the deep end! I did get out the car. Stan said to Jamie, "Look at her, man. You need to tell her the truth."

At first, Jamie refused. I said, "I know what you're all about, you jerk! It's over—get out of my life, and I don't ever want to see you again!"

For some strange reason, I followed my gut; I believed what Stan told me. He knew I was hurt. Finally we drove off. "You see, sweetheart? I would never ever do that to you. I'm not perfect, but I would do you right."

I looked in his eyes with pain, and said, "I believed you the first time; I just didn't want to hear it."

Since that night, Stan and I bonded. I developed a close relationship with Stan. I was still uneasy, because he was my brother Rodney's best friend. I spoke with Rodney about Stan, and he said our relationship wouldn't come between the two of them—you can't stop the way a person feels, so it was fine with him. As Stan and I got closer, months passed, and I started getting sick and sicker. I went to the doctor later to find out I was pregnant. My mother was right there with me when the doctor broke the news to me. He said, "You're pregnant."

My mother looked at me and said, "By who?"

I said, "By Stan."

She was just as shocked as I was, but from that point on she never turned her back on me. I remember this day so clearly—I knew I had to break the news to Stan. I started being around him less and less because I was getting sick all the time. When he came over, I felt that I had to end the relationship. Stan was very outgoing, and loved to go out. My brother Rodney and Stan always went out in my car. I was standing on the first floor of my grandmother and Aunt Betty's porch, leaning on the banister; Stan got out of the car looked at me. I didn't say a word to him.

Before I could open my mouth, he said, "You look like a ghost scared you; what's wrong?" I told him I just found out that I was pregnant, but he was not shocked at all. He looked at me and said, "Take care of my little girl, will you?" I was shocked—he didn't know what I was carrying.

I didn't think anything of it at the time. All I said was, "Where are you going?"

"Nowhere, I'll be there every step of the way for you and her."

"Okay, well—I feel that we should end the relationship, because I can't be there for you. I can't even go anywhere without vomiting somewhere, so you need to go on with your life, and I know you'll be there for your little girl, too!"

As time went on, I started getting very sick. I was still working at First National Bank, and finally I went to my boss. Before I could get a word out, he said, "I know you can't handle work right now—take a leave of absence, because you're a good worker and we don't want to lose you."

Before that, all my co-workers were very supportive of me; they even gave me a huge baby shower with lots of gifts for my daughter. She was blessed before I gave birth. My boss was very understanding, even though he was a man. I was impressed that he was so comforting to me. I left my job, and I was so sick that I went back to see my doctor. I said, "I know this is my first pregnancy, but why am I so sick?"

My doctor replied, "Your pregnancy is very rare—it's unusual that women get this sick. It's your body going through many changes. I advise you to take some time off work. I'll write a note stating you on bed rest need to be bed rest for the remainder of your pregnancy."

I told him I'd already had this conversation with my boss, and he understood as well; I had no problem with that. My pregnancy was getting so bad that every time I would open my mouth about anything, I would start to vomit! I ended my relationship with Stan; I was on bed rest and I couldn't give him any attention. Ladies, you know how that is—if you don't give a man attention, he'll go elsewhere, right? I didn't want to be selfish either, so I did what I had to do, but he would still come by and check on me. Most of the time I was sleeping. I was too weak to be bothered with anybody, even with female friends. My mother had to be the voice for me; I couldn't speak without vomiting. I would hardly get out of bed, only to shower, and that was hard. I could barely eat, or hold water down. I was losing a lot of weight.

My mother was getting nervous, so I went to the doctor

to check on my weight loss. The results came back. The doctor said, "Your baby is growing well. You are ninety-eight pounds, but the two of you are fine—your body can't handle the pregnancy, but you're holding on strong; therefore, you have to go home and stay on bed rest until your due date."

I was so upset that he could not find out why I was still so sick, and being upset made me feel even worse. So I went home the next day. A lot of my concerned friends came by the house. At the time I was in a group called "Family"—they were so supportive to me, but I had to stop singing, which they understood. My dear friend Angela came by from Connecticut. She looked at my teary eyes and said, "Look, my friend, you need to eat! I'm going to make you some soup and give you some water—you look dehydrated." She went in the kitchen with my mother, and my mother filled her in on what was going on with me. She came back in the room with soup and water; she actually fed me soup and water. I was too weak to hold the spoon in my mouth. I felt like an infant.

It hurt me that she was hurt looking at me and feeling my pain. She said, "I know you're hurting inside that you can't do for yourself." I felt relieved after that, because she could read what I was thinking without me telling her. Thank you, Angela! As time went on, I started to feel stronger, to the point I could get out of bed. My family decided to give me another baby shower at home. Another huge shower—my baby had anything and everything I could think of, even stocks and bonds from her godparents, because I had so much support. I had a lot of people wanting to be her godparents, so she ended up with three godmothers, and three godfathers—it was a hard choice.

I started feeling better. It was hard at first, but I accepted the fact that I was having a baby, so I started going back to church. Word got out all over that I was pregnant, but I didn't care what people thought about me. One Sunday afternoon we had an event at my church. We had out-of-town guests, and one of the bishops approached me after the service. He said, "I'm very disappointed in you; I have no more to say to you." He walked away, shaking his head.

His wife came up to me and asked, "What's wrong?" I told her, and she said, "No matter what, I still love you! You don't worry—he'll be all right!" She gave me a big hug and said, "You're special to him."

That wore on me for a while, but finally I said, "I'm having my baby and I don't care who thinks what about me anymore, nobody is perfect!" Later, I found out that the bishop had a child outside of marriage, Who was he to cast stones?

As time went on I started getting sick again. "Not again!" My aunt, Betty Ingram, and my grandmother, Ida Mathis, would always help my mother out with me. It was my first, and I didn't know what to expect, but they did. My grandmother said, "From now on you stay downstairs with me so we can help your mother look after you."

Staying with my grandmother, I felt so comfortable—she would turn on her TV, put on the gospel channel, and pray all the time. My aunt Betty would make sure I ate and got out the bed to get a little exercise. I was not happy about that, but she knew what she was doing; I knew my mother was drained... she had done so much supporting me and my pregnancy, but she needed a break. At one time she thought I was possessed...I just could not control being sick. I was a mess! One day I got up from the bed I started to think, *How can I have this baby, quick?* I couldn't take it anymore, so I went to the second floor (we had a three-floor apartment) and stood at the top of the stairs. My mind was playing tricks on me—it was saying, *You want to have that baby? Then throw yourself down the stairs.* I did just that—I went tumbling down the stairs, landing on the first-floor stairway.

My family ran out there and rushed me to the doctor. They explained what had happened. He took an ultrasound and ran all kinds of tests. Finally the results came back—of course I had a praying grandmother who knew how to get a prayer through. This was one time I was praying, sick or not. The doctor said, "The baby is just fine—I'm going to release her to go home, but she has to be supervised at this point."

I was only five months pregnant, and couldn't imagine getting through four more months. That was a long way to go for me. I wanted to have my baby much sooner than that, which led to my crazy thinking. When I got back home I went to lie down, with my grandmother. She said, "I'll keep an eye on her, she'll be just fine."Aunt Betty said, "Yep, she will." She just wanted to make me walk and get out of bed! That's Aunt Betty. Well, I finally slept like a baby off to sleep.

Mother Susie Mathis & father Theodore W. Mathis

Aunt Betty Ingram & Ida Louise Mathis

2. The Nightmare of Losing My Daughter's Father

Stan and my brother Rodney would always ask if they could borrow my car. *Why not?* I thought. I couldn't drive it; I was too sick to drive. Stan and I had already ended our relationship. We were on good terms, but we didn't date for a long time. However, he wanted to be there for me no matter what, especially when he found out I was carrying his daughter. I just didn't want to hold him back. Later that day, my brother came downstairs and asked if he could borrow my car keys to go out, I said no problem—and believe it or not, when I told him that, I didn't even vomit! Rodney said, "Stan and I are going to a party, and we'll be back later."

Again, I said, "No problem—just don't stay out too long with my car."

He chuckled and smiled, and ran upstairs to get the keys before I could get another word out! The two of them had thought my car was their car anyway! That evening I decided to stay downstairs with my grandmother. I've always felt like a baby around her—very peaceful, like I was getting better. The doctor told me the farther along I was, the better I would start to feel. I was getting there, finally. I drifted off to sleep that night. My grandmother had a police scanner in her bedroom; she kept it on all day and night, and it kept her updated about what was going on in the city of Boston. For some strange reason I wasn't in a deep sleep. It was kind of late, and all of a sudden we heard a very loud voice on the police scanner—one officer radioed another officer saying that two young men got shot at Leners Park in Dorchester. He believed that one had died at the scene, and the other was badly wounded but might still be alive—he wasn't sure.

A second later, I could hear my mother's phone ringing upstairs. All of a sudden I heard a loud scream saying, "Oh no, oh no!" I knew right then and there it was my brother Rodney and Stan. I said to myself, *I loaned them my car, and they never came back home.*

My grandmother jumped out of the bed. She said, "Oh, no—you stay right here. You have to think about the baby. Betty and I will see what's going on. Lie back down and breathe; I'll be back downstairs."

I said with a shivering voice, "Okay." I was in tears. They went upstairs. I started crying—I needed to let it out, but I couldn't cry loudly. I didn't want to do any damage to my health, I had to think about my daughter. I knew one of them was dead. I was so scared to death. If it was my little brother, I didn't know what I would do without him—he was my right hand. He would always look out for me in everything. We would sit up late at night talking about so many things. If it was my daughter's father, how was I going to explain to her that her father got killed before she was born?

My sister, Wilma Faye Mathis, ended up going to the scene of the crime. She found one of them on the floor of my car, as if he were crawling to get away to survive. She found my brother's shoes on the other side of the road. We got a phone call at the house, and my grandmother broke the news to me. She said, "Stan is dead, and Rodney is badly wounded." The next day I was told that my daughter's father was shipped off to Mississippi—and that was the end of that story for me. I never got a chance to say goodbye to Stan…only in my heart.

Later on that week I got the strength to go visit my brother at Boston City Hospital. In order to see him I had to pull myself together, for him and for my daughter. I walked into the hospital room. It was hard, seeing him lying there with bullet wounds in his leg. I knew he was in severe pain. He looked at me and said, "Don't worry—I'll be all right. I'll live." He said to me, "I have a lot to thank God for."

My brother came home a couple of days later; he limped into my room and said he would look after my daughter—he would stand in the gap for her father, and that meant being a father to her. Wow—I was so happy. As time went on, I knew I had to deal with the death of my daughter's father in my own way, but I felt that he was always there, like he had told me on the stairs, so I felt at peace with my brother around me. My pregnancy started getting stronger; I started getting anxious to have her here with me. I said, "If she has her father's eyes, I don't know what I'm going to do—I would be beside myself. Please God, give me strength—You know what's best."

My due date was fast approaching, and I started sleeping back upstairs in my own room.

3. Giving Birth

My pain started to get worse. I knew this wasn't right, but was not sure if they would send me back home or not. I had heard about going into the hospital in so much labor pain and the doctor would still send you home, pain and all; if it's not your time yet, then it's not your time…but who wants to hear that crap? Like my grandmother told me, that baby will come when he or she is ready to come out. I said to myself, *Oh, well—I endured the pain this long, and it's almost that time, so I might as well wait till my time.*

Before I knew it, my pain was starting to get worse. I went to the hospital. I felt like it was that time, but my doctors looked at me, and said, "Not yet, but it's getting closer."

Well, you know I didn't want to hear that. They said, "We are going to send her back home." I was hot like a firecracker! So we ended up going back home. I tried to go to sleep, but my pain started kicking in again. My grandmother came upstairs, pulled up a chair, and sat in front of me saying, "That baby will come when she's ready to come out."

Later that night my pain was unbearable—I couldn't even move. I started screaming so bad, my grandmother said, "I know you're getting closer; she will be ready to come in the morning."

Meanwhile, we sat up the entire night. I was in so much pain—it was ten times worse than a toothache. By the time I looked at the clock, it was 4:00 a.m. In the morning, my grandmother was still sitting up with me. It got so bad I started yelling. I don't know what I was saying, but she knew I was in pain. It was 6:30 a.m. I said, "Nanny, please take me to the hospital now—I can't take this anymore!"

My grandmother said, "Yep! Saundra Louise, you're having your baby today."

My mother and Aunt Betty got me dressed; my father said, "I'll carry her down the stairs."

My grandmother said, "I'll see you when you get back with the baby."

I chuckled a little and looked at her as if to say, "I'm not coming home until I deliver, whether the doctors like it or not!" That was my pain talking.

I finally got to the hospital and was seen right away. The doctor said, "Yes, she is ready—looks like you've been in labor all night long, at least thirty-six hours." I was screaming. He said, "The more painful it is, the quicker your baby will come."

I said, "I can't bear this pain anymore."

So he said, "Lie on your side and don't move."

I yelled and said, "Hurry up!" They gave me an injection to ease the pain. Before I knew it, I was numb, and the doctor told me to start pushing. After a couple of pushes, she came right out. What a relief! My pain was all over—I felt no more pain; that was so weird.

My daughter was born on December 30, 1986

At 8:24 a.m. The whole time, my mother was with me in the room; after the doctors took her, my mother left me on the other side of the room, holding her and kissing her, saying, "Look at my little angel! Look at those big eyes!"

The doctors and everybody were on the other side of the room; meanwhile I was saying, "Hello, I'm over here! I am her mother!" I guess they'd dealt with me long enough! I turned over and said, "Oh well—I need some sleep now." I knew my

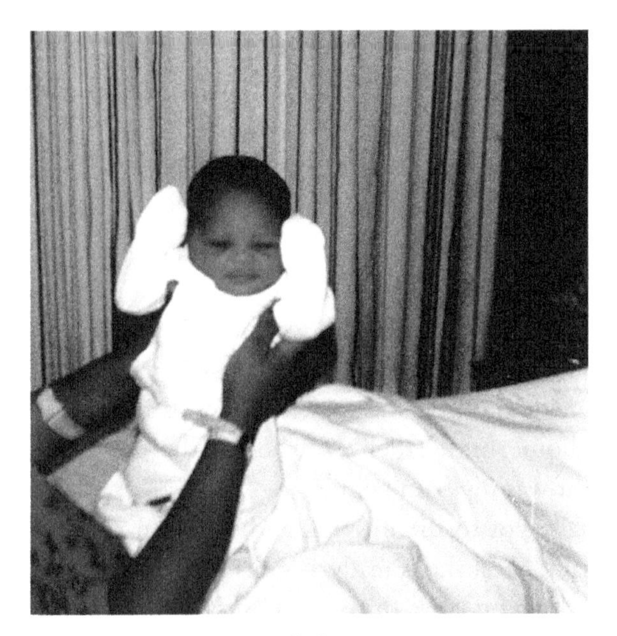

Anisia's baby picture

daughter was in good hands, because my mother wasn't leaving the hospital! The next thing you know, I was out. My mother didn't leave my sight. I stayed three days in the hospital. My Aunt Jenny and my mother came back to take me home. My daughter had been blessed with two baby showers, and when I got home she had more gifts and brand-name clothes. When it came to my daughter, she didn't have to want for anything—she had everything.

At the time, I did not understand why it was so hard for me to bear this pregnancy. Well, I know why now—my daughter is a "blessed child," heaven-sent from God. My brother Rodney kept his word; when she got home, he took over. He even gave her a nickname: "Littleone." My mother was her second mother and my brother was her father. I would get up in the middle of the night and look around; if she wasn't in my bed, I would go into Rodney's room and she would be all cuddled up next to him. I would laugh and say, "Oh, I'll leave her there," and go back to bed.

My brother Rodney was a drummer in the church. He would always take my Anisia, set her on his lap, give her a pair of drumsticks, and hold her while he was playing the drums, and she would sit and steer at his hands while he was playing. She wouldn't miss a beat—she didn't budge, nor did she show any emotions. I asked myself whether she liked it or not—it was hard to tell. She wasn't emotional at all, but she didn't want him to put her down off his lap. He would take her for walks as she got a little older, and she would ask him to buy her drumsticks. He did just that. Later, she started taking my mother's pots and pans, beating on everything she could find—she would make such loud noise, I knew I was in trouble, but my mother would allow it.

brother Rodney Mathis & Anisisa

4. Why Anisia Is a Blessed Child, and How Her Musical Life Began

Anisia is a musician; she started playing drums at the age of thirteen. Even though she was focused on playing the drums, she managed to be an A student in school. She was very quiet and always kept to herself, but she never missed a beat. She started playing the drums at the New Life Restoration Temple Church, in Dorchester, Massachusetts, for Pastor Bishop Bernard N. Bragg. One Sunday she was playing the drums at a service and we had a special guest there, a particular young lady from Baltimore, Maryland. This young woman spoke into my daughter's life. She said, "Look out, Boston—your daughter will be the next Sheila E. She will be known all over the world through her gifts and talents. Many doors will open for her, because she's so humble and gifted."

Later, she participated in a Mars drum store contest for drummers in Boston. It was sponsored by Sheila E. and The Little Bunny Foundation. Her team interviewed a huge number of participants from all over the world. They had to select six drummers—and yes, my daughter was one of them. On November 21, 2001 they flew Anisia, along with David and me to LA, California; she performed live and recorded live onstage with Sheila E., the five other participants, and Sheila E.'s family. I was one proud parent in that audience. Later on Sunday morning she was playing at church for the worship service, and we had two guests from Berklee College of Music; they were in charge of the mentoring program at Berklee. After the service, they approached me and said, "Wow! She's bad; we would love to have her participate in our program for the summer." At the time, she was only in the 7th grade; I was a little nervous, but I knew this was what she needed. Later she attended the System 5 summer program at Berklee College—yes, she ended up going there every summer till she graduated from high school. After she graduated, my husband got a job offer to relocate to Houston, Texas. I knew she was going off to college; therefore we ended up moving once she was settled.

She received a full scholarship at the Berklee College of Music, and she received several endorsements from different music stores. She played for a tribute band along with seven world-famous drum performers onstage all at one time. The performance was held at the Berklee College Performance Center. What a performance—I'll never forget that night. Anisia always wanted her own production company; also, she loves kids and they love her too. She always said she wanted a youth center for kids to teach music, and I've always believed she would end up running a school one day in the near future. Well, when she was about to graduate from high school, she toured in Germany along with other students to teach music—they would allow her to travel through the schools for different events. Finally she graduated from Berklee College of music on May 9, 2009.

She attended the Revival Time Flame of Fire Church—she really loved it there, and she wore many hats. She became the minister of music, and also worked the sound system. By 2011, she is the community music director of the Boys & Girls Club. While I was away in Texas, my daughter "adopted" parents—two special married couples who mentored her. They played such a big part in her life; they took her under their wings: Angela E. Frazier-Atkins and her husband Kevin A. Atkins. She was the one who fed me when I was pregnant with Anisia. I really thank God for the both of them being such a big part in both of our lives. The two other special godparents who took her under their wings are Latrelle Pinckney-Chase and her husband Tyrone Chase. To this day, they both still look after Anisia. I thank God for them, and for my grandmother, Ida Louise Mathis; she's gone on to be with the Lord. I am thankful for my aunt, Betty Ingram, my brother Rodney, and the inseparable parents Theodore W. and Susie Mathis. Now I can see clearly why I had such a rough pregnancy—I gave birth to a blessed, gifted, and talented young lady.

daughter Anisia

We'll forever miss you Stan, R.I.P.!

5. I Met My First Husband—Love at First Sight

I really loved to sing, but I'd always wanted a role in a play. I met a young lady named Kami Wilkins, who was the director of a production. She had just written a play, and was looking for cast members; she heard me sing at an event, then she approached me and asked if I would be interested in singing and acting in her skit. I was honored to do it—without hesitation I said yes, I would love to do it. Kami said there was a young man she had just met; he lived next door to her, and she found out he loved to dance and he was very artistic! She said he would fit right in what she was trying to do. Kami said, "Girl, I have the perfect role for you and him. You have got to meet him, he's good-looking and has deep dimples."

I said, "Okay!" I thought nothing of it.

She said, "Look, I'll pick you up for rehearsal—you can meet him then." Well, the following week came, and she pulled up in front of my house, tooting her horn. I had just walked in the house. Needless to say, I was starving; my mother had just finished cooking dinner. I grabbed some chicken and ran down the stairs, eating at the same time as I was walking toward the car.

I was going to jump in the back seat of the car when I noticed that someone was sitting back there. Kami said, "Come sit up front."

The young man in the back seat kept smiling. I said, "Please forgive me for eating like a pig, but I'm hungry—if I don't eat,

I can't function in the rehearsal tonight." I was so embarrassed, eating in front of him. I was smacking and talking at the same time!

Kami said, "This is the young man I was telling you about." She introduced us. He kept smiling with his chinky eyes and deep dimples. In my mind I said, *Maybe he's laughing at me—oh, well.* I kept eating my chicken.

Finally we got to the rehearsal, and everyone was great. After rehearsal, Michael approached me and asked if we could exchange phone numbers. He said, "I live around the corner from you."

I said, "Wow! That's close."

When I got home, as I was walking toward my bedroom I could hear my phone ringing—yes, it was him. I asked him, "Why were you laughing at me?"

He said, "No, I wasn't laughing at you; I was amazed when I saw you. I thought you looked sexy—when I saw you, you blew me away watching you. I'm an artist and I love to draw…I was imagining drawing you eating."

I laughed at Michael and said, "Okay, at least it wasn't personal."

From that point on we talked on the phone every single day and night. We started dating, and he fell in love with my daughter Anisia; it made me grow closer to him he treated her well, and he really loved her like his very own daughter.

6. Moving Out of My Parents' House

I moved out my parents' house at the age of twenty-three. I was so excited to become totally independent on my own—wow, my own apartment was big for me, and back in the day that was big for us. I had a very good job; I worked at MaComber Construction Co., where I was the accounts payable and receivable clerk in the office on Atlantic Avenue in Boston, near the waterfront. I managed to purchase my second car, my sedan… it was very small, but it got me around. I found a beautiful apartment near Wollaston Beach in Quincy, Massachusetts. This was my first apartment. I finally got settled in my apartment along with my daughter. Anisia was so excited about living near the beach. Michael would come by and walk her to the beach—they would play in the sand and throw rocks in the water. What got me was that when he brought her home, she would come with a bag full of shells and dirt! She would have all the sand trailing on my floor—that drove me crazy! Hey, I was a neat freak. I laughed and told myself to get over it; it wasn't that serious.

I asked my daughter how she felt about Michael. She said, "I really like him; he's nice." She never said much anyway, so I knew she would tell me the truth.

Michael was so romantic—I can remember when he asked if we could have dinner on top of the roof of my apartment. I laughed and said, "Are you nuts?" But that was his style. We fell deeply in love. Finally he brought me by his house to meet his parents—he really wanted me to meet his mother. He said we had some things in common. I met his family, and we bonded as time went on. I've never had a problem with his family, and they never had a problem with me and Anisia— they fell in love with her too! My family loved Michael and his charming ways; he loved to smile all the time. He had a special bond with my family—he was just a likable guy. Everybody loved him.

7. Michael's Proposal

Michael and I would be around each other so much that we were like glue sticks jelled together. One evening Michael came by my apartment and we decided to take a walk. On our way walking toward the back of my apartment holding hands, he looked at me and said, "I can see you being my wife one day." I looked at him with gleamy eyes like I wanted to cry. Michael said, "Saundra, you're everything a man could ever want."

Well, I had always done everything on my own and had my own mind, and my own way of doing things; even as a child I was very independent—that's what my parents had instilled in me. I would cook; I was obsessed with cleaning; I loved fashion, singing, and people; I loved to enjoy life—yes, I loved to have a good laugh every now and then.

Michael and I drew closer and closer to each other, as if we could not be away from each other, and as time went on I felt it was almost that time. Michael and I got married on April 28, 1990. We shared some good times together—he would bring me flowers, little gifts, maybe a drawing…he loved to draw, and he would make things with his hands. He supported me in my music career—if I had an engagement, he would be there. Wow, a man after my own heart! He would show me off to his friends with no shame about anything—that's just the way he was. While he was drawing and sipping his coolers, I would just sit there and watch him draw; it was so romantic.

He would always get phone calls in the middle of the night from his friends either to meet them outside or to hang out—

that used to bother me. I was still deeply in the church, and it was like we both were trying to step out into each other's world just to meet halfway, without condemning each other. Later, things started to get a little tense. I was pregnant, and I would say to myself that since I was pregnant and the family was expanding, we needed to get a bigger place. My parents knew something was wrong, because I was too quiet—the one thing about me was that I would always show something was wrong quicker than telling it. Since I was moving and packing, and not only that but also trying to keep the marriage together, my mother asked if she could look after Anisia until we got ourselves situated, but they would never interfere in my marriage problems—my parents would always say that was between me and my husband unless something happened that made it necessary for them to step in, but that never happened.

Our marriage started getting complicated. I was also getting sick the second time around—my body just couldn't handle being pregnant at all. I was always sick day after day and night after night again. I even had to take another leave of absence from my job until I got better. I was told men get morning sickness too, so maybe it was affecting him as well. I said that could be our problem—we were going through the emotions together.

8. The Loss of My Younger Brother Rodney

One day when I was lying down, a friend called me up on the phone and told me to talk to my younger brother Rodney. I asked why, and all she said was that I should tell him to watch his back. Obviously she knew something. I called him and told him just that. He said, "I'm not worried about it—it's just talk."

I said, "Okay, I'm going to lie down."

I didn't feel comfortable, so I asked Michael to take me by my parents' house, and said I would just hang around there, so he dropped me off late that afternoon. All I can remember is that my car was parked across the street directly across from my parents' house. Later that night I was lying downstairs on the first floor of my Aunt Betty and grandmother's house when my parents got a phone call from the neighbor across the street, saying, "Go outside—Rodney is laying on the ground!"

Everybody started screaming. I couldn't get up quick enough—I was too sick too and weak again at five months pregnant. My sister ran outside. My brother Rodney was lying in front underneath my car, with lots of blood under the back of his head. My sister lifted his head up and asked what happened, but he couldn't even respond. It was another nightmare—he was rushed to the hospital. When he got there, we stayed all night long. During that week, the doctors called the family together—you know what that means. We had to make a big decision: either keep him on life support, or pull the plug—what a huge decision, to let him go. It wasn't the same without him. What a tragedy. I felt now I had lost a brother and a father to my daughter Anisia. Later I had a dream about my brother; he was walking toward me…I was not sure whose house we were at, but he was walking toward me, limping very slowly. I had my daughter Anisia standing next to me. He grabbed her and set her on a shelf, and then he looked at me, and looked at her; he never said a word to me, so I said, "You want me to take care of her, and I know you'll always look after her." Then he nodded his head and walked away, limping very slowly. I woke up out of my dream, so I was at peace again, that she would forever be watched over by her biological father and my brother Rodney.

Rest in peace, my little brother Rodney Mathis; I will always have you in my heart deeply, and will forever miss you always love you.

I had experienced two deaths at five months pregnant, and I had to get rid of both of my cars, because the police needed them for evidence—not only that, but there was blood in and on both cars, and I know I could not have handled it.

Michael started working at a good job. He mentioned to his boss that I was pregnant, and we needed to find a bigger place to live. His boss suggested that we rent his two family houses in Somerville, Massachusetts. Not only that, he gave Michael a car—that was two blessings in one. Michael gathered up some of his friends to help us out, because I was seven months pregnant. We finally moved to Somerville. We moved into the second floor of the house; it was very roomy inside. The attic was huge. I would hardly go upstairs, because I was huge and had too much weight to carry. But it was lots of space. I was happy I had a room for Anisia and my new baby girl, even though she wasn't born yet. I managed to pull myself together after the loss of my brother; I said, "God is keeping me in my right mind for a reason." My job was always good to me. I was still working at Macomber Construction Company on Atlantic Avenue, and my co-workers and boss were very supportive of me. I got very close to one of my co-workers, Dorothy, who would help me out with anything I needed. She also wanted to be Kyva's godmother, and that was special.

9. A Painful Time When I Wished I Had Rodney's Advice

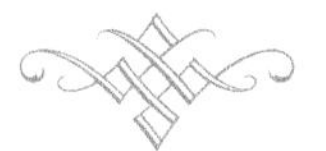

Michael started coming home very late—sometimes he would come home the next day. I always confided in Dorothy, because she would listen to me and give me her support in any way she could. I wasn't going to put up with Michael's behavior; I was the type of woman who was not afraid to ask questions. Some women would be intimidated, because they're afraid of losing their man, but not me—I had very low tolerance for misbehavior and being disrespected in a marriage. I can remember it was getting closer to my due date. It seemed like he would hang out even more instead of being around to make sure I was all right, but I stayed home doing the wifely duties, as usual.

On November 13th, 1990 his best friend Choy came by the house to hang out with him. We struck up a conversation. He looked at me and said, "I know what you are going through—I can see it in your eyes, and I know him very well too. You don't deserve this kind of treatment. Why is he not at home with you, knowing that you will be having your baby any day now?"

I said, "Choy, I'm tired of it, and I'm not the type of person to let any man take advantage of me, but right now I have to focus on having my baby."

It was getting very late, and we talked and talked; when I looked at the clock it was 1:00 in the morning. I said to Choy, "Thank you for staying up all night listening to me; I needed to vent my frustrations."

I asked him to go home and get some sleep. I told him I

would be all right, and that I would call if anything changed. Choy gave me his phone number and said, "Please keep me informed—call me if you need me."

Finally, he left. I was pacing the floor, heated like nobody's business. I kept looking at the clock on the kitchen wall. Michael did not call to check on me at all. I started cleaning doing things around the house just to keep busy. I was so angry at Michael for not calling or showing any concern—it was 3:30 a.m. and there had been no sign of Michael.

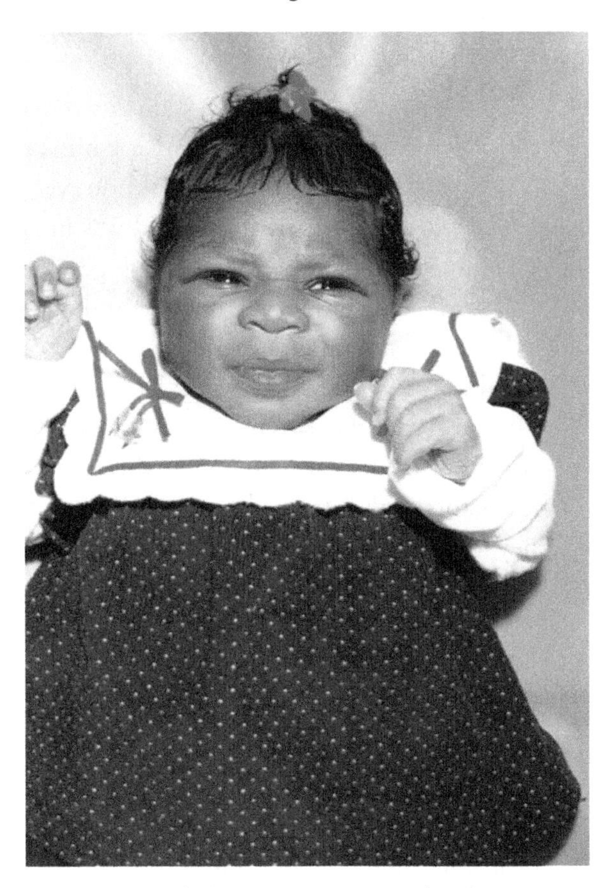

Kyva's baby picture, youngest daughter

I started asking God, "WHY ME!" Look at all the things I'd lost in the past few years of my life—two loved ones, and now I didn't have that man to woman advice from my brother. He would always have his input when it came to men in my life. I started crying more and more, until I couldn't cry anymore! Everything started wearing on my mind. I laid down for a second. I looked at the clock; it was around 6:00 the next morning when I got up, took a shower, got dressed, and started pacing the floor again. I was so pissed that Michael still was not home. I started feeling pain—it was getting worse and worse. I started holding my stomach.

Michael walked in around 7:00. I was hurting so bad that I couldn't even say anything to him—I did not ask him any questions; I couldn't open my mouth. All I could do was scream, "I'm in pain!" He knew right then and there I was in labor.

Michael screamed, "Oh, my God—let me get you to the hospital right away!" He grabbed my coat and wrapped it around me. He picked me up and carried me to the car. I was starting to bleed a lot; I believe my water had already broken by the time I arrived at the Brigham & Women's Hospital. Michael jumped out of the car, grabbed a wheelchair, put me in it, and the doctors took it from there.

When they looked at me they said, "Wow—you must have been in labor all night…your baby's head is coming out…we can see her." They said, "There's no time to give you an injection to numb you—sorry, you have to go natural all the way."

I was screaming, telling them to hurry up. Michael started getting nervous for me. They rushed me onto the table, and all the doctors came running into the room, saying, "She's ready—hurry, her head is coming out!"

Michael squeezed my hand and said, "You can do it—just breathe." The Doctor gave Michael instructions on what to do—I believe it took just two good pushes and there she was. Michael cut the umbilical cord. They said, "It's a girl!" which we knew already. She was born on November 15, 1990 at 8:24 a.m. at the Brigham & Women's Hospital. Wow—all I could think was *This is it for me!* The weird thing was that right after she came out, my pain was gone—what a relief! Michael gave her the name Kyva Myshan.

I yelled and told the doctors, "Hey, go ahead and tie my tubes—I can't handle being pregnant again. Doctors, you don't understand the things I had to deal with, so please do it."

They looked at Michael, and said, "You both need to go home and talk about it."

I said, "It's my body, not his—he doesn't have anything to worry about—give me a break!"

The doctors said, "Sorry, it has to be a mutual agreement between the two of you. You can't make this decision by yourself; that's not fair."

I said, "I know both pregnancies weren't easy for me—I had it in my head that when I was at five months something would happen, and I don't ever want to go through that trauma ever again!"

I spoke with Michael a month later about going back to have my tubes tied. He understood me, and he didn't want to see me suffer anymore. We went to the doctor, and the doctor suggested that if I wanted to get my tubes tied at that time, they had this special clamp that could be reversed, if I changed my mind later in life. Michael was comfortable with that, and the doctor said if I wished to untie my tubes, there might be a chance that I couldn't get pregnant again. That was a chance I would have to take. I went into my marriage long-term, till death do us part—at least, those were the vows we took, but seeing all the problems we went through while I was pregnant,

I said maybe we both were going through pregnancy changes. I don't know; I thought the baby would draw us closer together. Ladies and girls, you can't change a man if he doesn't want to be changed—take it from somebody who knows what she's talking about. Well, I went to the doctor, and I said, "I have to do what I have to do." That was over…if it was God's will for me to have another child, then it would will happen, and till then, God knew my heart couldn't bear it anymore. That was the attitude I took.

As time went on, Michael started becoming distant from me, but he loved his daughter. He would still get calls in the middle of the night, or he would get up and go out, telling me he'd be back. I didn't buy it anyway—I would then give him a mouthful. He was not stable when it came to the marriage; he started acting weird to me. But that was just Michael; he was his own kind of guy. Two months after Kyva was born, Michael left us there. I saw that I couldn't handle the bills by myself anymore. I was living in his boss' home. I sat down and spoke with his boss, who was also our landlord.

He said, "I don't know what's wrong with him, but you need to do what you need to do. You and your two daughters are welcome to stay as long as you need to."

I didn't want to take advantage of him— that's not how I was raised. My parents taught me well. I started confiding to Dorothy. I still had a good job to go to, but the bills were hard. I didn't have a car, because I got rid of it. I spoke with my parents and told them what I was dealing with. My father said his best friend was selling his car, and he thought it would be reasonable—he'd look into it for me. The next thing I knew, my father purchased the car for me, and I made sure I paid him back. It was an LTD Ford—a very big car for little ole me! You couldn't tell me anything. I said to my parents that I needed to move—I couldn't handle my bills. Not only that, but I had one daughter in elementary school, and a newborn. My mother said, "Don't worry—we will watch Anisia for you. Her bus stop is near the house, so I'll walk her back and forth."

My mother and father kept their word. Dorothy asked me and Kyva to come live with her. It was kind of far on the outskirts of Boston, but I had no choice. I called my father and he said to let him know my moving date—he and his best friend would come help me move. My mother offered to help me pack everything up, so I wouldn't have to do it by myself. Still no Michael—at that time I didn't even care anymore, since I felt he didn't give a damn about me and his daughter. Dorothy knew I was struggling trying make ends meet, and she knew how much my paycheck was; we both did payroll together. Michael was not helping financially—that stopped altogether. At the time I felt like I wanted to get away from everything, just to get a break! I needed to clear my head.

10. Damage!

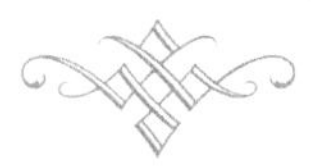

Dorothy's mother lived in Norwell. She offered to babysit my youngest daughter Kyva at no charge, but I would offer to give her something, even though I didn't know her mom that well—since she'd kept other kids, I felt comfortable around her. She was very close to her daughter Dorothy, so that said a whole lot. My father and his friend came by to help me load my stuff on the truck. Yes, they moved me all the way to Situate, Massachusetts. Dorothy said I could store all my things in her basement. The only things I needed were my two daughters' bunk beds for them to sleep on, and my clothes—everything else went in the basement. I've always had big furniture and a lot of clothes; I had very nice things all the time.

Dorothy gave us our own room, and said, "Your older daughter is welcome to come live with us too."

I managed to go back and forth to Boston to see about my older daughter; I made sure she had money for school, and to check up on her with the school. I was always in her life—my mother made sure of that. I even sometimes brought her to Situate on some weekends, but she didn't want to be away from my mother. My parents took very good care of Anisia; I knew she would always be in good hands, especially with the angels looking over her—my brother Rodney, and her father Stan. Finally I got settled in with Dorothy. Dorothy's boyfriend, Tiger, would always prepare dinner for us—this was somebody who knew how to make a mean pot roast! That was our favorite. They both were Caucasian, and knew how to cook—that was up my alley! I never had to cook or lift a finger living with Dorothy. She lived right off the beach and it was beautiful living with her—what peace of mind!

Michael would come out to visit us every now and then. Yes, I still missed him. Love doesn't go away that easy—well, true love doesn't go away that easy. I gave him my heart and soul—we were soul mates.

One day it was raining so hard that Dorothy's basement got flooded out. The next day, Dorothy and her boyfriend Tiger went down in the basement to check on things. She came back upstairs, looked at me with tears in her eyes, and said, "I'm so sorry—it looks like all your things got damaged." She was right; everything was moldy, and there was so much water in her basement that we had to wear boots in the summer time. She was flooded out—everything I had was lost. I had even put away my daughter Anisia's saving bonds that her godparents Ann Davis and Joseph Davis gave to her just about every Christmas, but I was holding on to them in a special place and didn't realize till later that they were packed away in the basement with everything else. I was hurt, but the way I look at it—look at all the things I've lost so far. These are only material things; compared to life, this loss couldn't measure up. I can always get material things back, but I can't get back my loved ones. I accepted what had happened, and I knew that I would have to start from scratch again when I moved.

Oh well, I thought, *what can I do?* Dorothy felt so bad she wanted me to file a suit against her landlord, but I told her, "Don't worry about it—you gave me a roof over my head, along with Kyva; I can't ask for anything more than that, so let's move on, okay?"

She cried and cried, so the following week I put some things on layaway just to reassure her I'd be okay! She was happy, and she paid for some of my things. I was ever so grateful that she allowed us to stay with her—she didn't want me to give her money for anything. She would not take it, and neither would her mom let me pay her for babysitting Kyva. They would not take a dime from me. Everything I had was mildewed; therefore I was looking to start my life all over again.

So much had gone wrong in my life. I was commuting back and forth from Situate to Boston, which started taking its toll on me and on my vehicle; therefore once I got myself together I knew I had to move back to Boston—I knew I had to put my daughter in daycare sooner rather than later; she was very active and easily bored. Michael's mother offered to let me live with her and her husband; they are the greatest people I've ever met in my life. They always took care of Kyva and helped me with anything I needed for her and Anisia; they would help me out in any way they could. So I felt good to know I would always have their support on my side. They did what was best for their granddaughter, and took me as their very own daughter—now that's love. I'm ever so grateful for his family, who offered to let me to come live with them till I got on my feet. I stayed with them for a couple of weeks—maybe a month.

Finally, one of their family members rented out the first floor of the house, and I knew I had to move. I still had feelings for Michael; he was coming over back and forth. He had moved out with a friend of his, but I was living with his parents, even though I knew it was temporary. Finally that apartment came through for me. I still had my Ford, which was still running well. Another friend of mine offered to babysit Kyva till I got settled, but my goal was to put her in daycare. I moved into a two-bedroom apartment. My sister in-law Trenda lived upstairs, so I was happy with that—she stuck by my side, too!

She would look after her niece all the time; she would even braid her hair for me from time to time. I loved that. Kyva also loved her husband Pete—he spoiled her. You see, I was surrounded with so much love in spite of what I had to go through. Michael's mom taught me so much while I was living with her. I was in love with her cooking—yes, she could throw down! I tried to copy her cooking every now and then.

Finally we moved, because it was very difficult to see Michael come and go; I still loved him. He broke the news to me that it was over. Well, I had figured that out a long time ago. He said he didn't mean for it to happen this way; he wasn't ready to settle down. No kidding—I could see that! I never took him to court, because his family was always there for me, so my daughter was never without; God always sent me help for her...that was a blessing from above. For some reason it was hard for me to let go my feelings for Michael. I even asked God to take away everything I had for him, but it just didn't happen that way. It took some time, and then more time. It was pretty obvious that he was seeing someone else, but I never knew who. He still would come by the house to see his daughter, or maybe sometimes to take her to the babysitter's house, or her doctor's appointment.

One day he came to the house to drop our daughter off; I got in his car on the passenger's side, and there was a picture of a Caucasian girl on his rearview mirror. I lost it on him, I got out the car and hit him—he shoved me up against the car and said, "Go in the house." I hit him again, but I knew I was provoking him. For the sake of my daughter, I stopped—she loved her dad. That was all I thought about. I was told that one day when he came by the house, he didn't come inside, but I went to the door—he had another woman with him, and boy did I lose it! He left so quick. I took one of my high-heeled spike shoes and threw it at them down the street. (That's what I was told—I don't remember.)

The last thing that sent me over the edge with Michael was I found out he was dating someone else—Kyva tried to tell me her name, so I confronted Michael about it, and when he told me who she was, it sent me over the edge. The woman he was dating had a brother, and he was the one that killed my brother Rodney. I held a long grudge against him for that—no, it wasn't her fault, but it was still fresh in my head; it took my family some time to deal with it, and she just reminded me of what happened. He didn't date her for long, but it took me some time to forgive…I'm over it now. I had to keep telling myself, "Move on—you deserve much better. Give yourself some time to get yourself together. Who doesn't want that special person in their life, and to feel loved?" But I had to accept things for myself and face the truth. I did end up moving into a two-bedroom apartment, and got back on my feet.

My dear friend had a nice daycare. I put Kyva in her daycare and got settled at last—she offered to watch my daughter in order to cut down expenses, but I'd had already had someone in mind. I still managed to keep my job…I worked at the Le Meridien Hotel in Boston. I loved that job! I worked the overnight shift. I was my own boss. Yes, I loved it. Well, things started working out for me, but I allowed Michael back in my life again. It wasn't right and I knew it—he still went out at night, going back and forth. I was still pursuing my singing career—I was in a play at the time. I met a well-known actor (I can't mention his name) and we had an event at Martha's Vineyard. This actor and I started chatting about life in general. He asked for my phone number and I gave it to him, trying to get over Michael. He said, "I can help you get back on track—I can feel and see that you've been through a lot." Well, the next day he called me and said, "Look, Saundra, I just got a call—I'm in a limousine on my way to the airport. I need to fly back to California asap! Would you like to come along, along with your two daughters? I'll take care of all of you."

Wow—that sounded too good to be true, but I didn't know what to believe anymore. It blew me away. While I was talking to him, my phone beeped. I clicked over to the other line, and it was Michael. I was trying to explain briefly to Michael a decision I was going to make—of course he said, "Don't do it; I wouldn't do that if I were you."

11. Looking for Love in All the Wrong Places

My head was going in all different directions—"He loves me, he loves me not"—that's what was going on in my head. Well, ladies, you know how that is, when you still have feelings left inside of you. He talked me out of it, which didn't take much, as you can see. I clicked back over the phone to the actor and said, "Sorry, I can't just pack up and leave. You can go on—I do have a lot of things to consider, such as my two daughters."

He said, "Okay—if you ever need me, or just want to talk, maybe to vent out about anything, please don't hesitate to call me. Got to go now."

He hung up the phone. Later the doorbell rang—it was Michael. I was so confused, because I knew Michael wasn't stable with our relationship. I said to myself, *You are a fool, girl—why did you do that?* Later that night, Michael was watching TV and the doorbell rang; he looked out the window and said, "Don't answer it," but the person kept ringing the doorbell. I refused to let them keep pounding on my bell—it was so annoying. He said, "Say I'm not here." I looked at him as if to say, *What are you into now?* I did answer the door. This guy was dressed in all black—black hoodie and black jeans; he had his face covered, which freaked me out, but I wasn't afraid of him. He had red fiery eyes—he looked like the devil! It looked like he was carrying a weapon, but I was not scared—more upset.

I said, "Can I help you?"

He said, "Where is Michael?"

I said, "He's not here. What do you want with him?" He looked at me very strangely, as if he did not believe me, but something told me not to say anything. *Think about your daughter in the house—don't push it.* I said, "Look, he's not here."

He said, "Okay—tell him I'm looking for him."

I said, "Okay, I'll tell him."

I closed the door and waited a few seconds. I approached Michael, and boy did I let him have it. Michael said, "Don't worry about it; I have it all under control."

"Control, my behind!" I said. "You're putting our lives in danger!"

But he never told me what it was about, and at that point I didn't want to know. He ended up going back out that night. I didn't know if he was coming back dead or alive. As time went on, I hadn't seen him for a couple of days—we were still married, but this was it; I'd had enough, and whether he returned or not I wouldn't ever trust him again, so I threw the key away for good, but he never came back anyway. I knew it was totally over between him and me; I just got that feeling, when you come to a realization and have to accept reality for what it is. I said to myself, *Girl, you really have to move on and mean it this time—you're looking for love in all the wrong places. Think about your daughters' lives, not just your life. This is not good for the two of them, either.* One day I was sitting on my living room floor, along with my younger daughter Kyva, and so many things started going through my mind. I felt this hurt and anger inside of me—all I could think about was that I wanted to hurt him like he had hurt me. I called my friend Cheryl and told her that I was going down to his job to destroy him the way he had destroyed my life.

Cheryl Hawkins-Coke, girlfriend

12. I Appreciate You

Cheryl sensed that it was getting serious. We hung up the phone, I got dressed to go down to his job, my doorbell bell rang, and it was Cheryl. I was already dressed, ready to go. I asked her to look after Kyva for me until I got back. She said, "What do you think you're doing—and where are you going?"

I said, "I'm going to his job."

Yes, I was way out of control—I allowed the enemy inside of me to talk to my mind. It was so bad I believe she slapped me, and said, "Pull yourself together!" She grabbed me so tight. She said, "Think about your two lovely beautiful daughters to live for—it's not just you. You have to stop and think clearly. It's not worth it, girl."

I started crying. She grabbed me again and started crying with me. Yes, she talked until I understood what she was saying to me. I took a good look at my daughter Kyva, and I started crying. Cheryl decided to sleep over—mind you, she had her own family. She had a son to look after, but she was willing to stay with us; she felt that maybe it wasn't out of my system that quick. I cried myself to sleep on my black leather sofa. I woke up the next morning. I had slept on the sofa, and Cheryl and Kyva slept on the love seat. Cheryl asked if I were okay—it was like a nightmare for me.

I said, "Yes, I'm okay, and thank you so much!"

She said, "Are you sure you're okay? Because you lost it for a minute—no lie, I was nervous, girl!"

I chuckled and said, "I'm sorry, but I'm tired of feeling this way—I felt helpless."

She said, "I'm glad you came to your senses…it's not worth it. And don't ever scare me like that again, you hear me?"

Cheryl would always encourage me; I totally relied on her for everything, because I trusted her so much. Whenever I needed a babysitter she was there for Kyva; she would always come by the house to take her off my hands. Kyva was a handful, and she loved to stay busy, like her Aunt Cheryl.

I found myself never having any money, living from paycheck to paycheck, trying to make do with what little I had. I felt that I could never get ahead in my life, and never had extra money for myself. But I always managed to dress well with what I had, and kept my two daughters looking well; I kept a good clean apartment all the time, and made sure they ate well. There were times when I was so low on food that I would call Cheryl and ask if I could borrow money to go grocery shopping. Cheryl never asked for it back, and never complained, she would come by the house with bags of food, open my refrigerator up, and start loading up food.

I would ask, "Where did you shop that quick?" and she would say, "At my house—I just gathered up something I thought you could eat. Don't worry about it…I'm doing this from my heart, and I don't want anything back from you. I'm here for you; that's all that matters."

Of course I started crying and she started crying; meanwhile, Kyva was looking at us both like we were crazy. I always felt she understood me, and always knew what I was going through—believe me, we were there for each other at times, but she was always there for me no matter what I was going through.

13. What's the Use of Living—to Heck with Life!

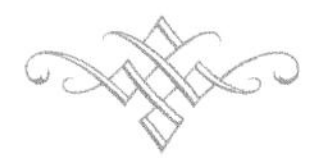

Cheryl never judged me or looked down on me. Most of all, she treated Kyva like her very own daughter that she didn't have yet. To this day, she still treats Kyva the same way—like a daughter. I knew people were telling her to back off, that it was too much, that she was doing way too much for Kyva—but she didn't listen to them. Real friends will be there for you no matter what. I seemed to talk to God a whole lot—I asked Him what and where did I go wrong in my life, and why people can't appreciate me for who I am without hurting or taking advantage of me. I seemed always to find time to spend in my living room—it was my resting place where I found comfort. I started thinking about my life again, and the mistakes I had made. I felt that I was a failure; I wasn't good enough for anybody. I said, "My life is empty and over—God is disappointed in me." I've always believed that God would put no more on me than I could bear; that's what He said in his word, but burdens started feeling heavy. I said, "What's the use? It's not worth living anymore. I can't seem to get ahead." I even told myself that Cheryl would look after Kyva and my parents would look after Anisia; they would be all set.

I just felt bad for myself, like I wanted to end my life right then and there. This was all going through my head, and meanwhile Kyva was sitting on the sofa looking at the TV. I felt I couldn't control the way I was thinking; I couldn't even call on God to help me. I felt I couldn't open my mouth to call on

Him to please help—I had to let my mind talk for me. The next thing I knew, I was eating poison— I wanted it over, I kept eating it over and over. My phone rang. I ignored it at first. It kept ringing and ringing. Finally I answered it—it was a dear friend of mine named Adora. Kyva had fallen asleep. I picked up the phone and started crying. Adora said, "What's wrong?" I told her I didn't want to live anymore, and she came over to the house right away. She didn't live far at all. She called my mother and father.

Adora explained to them what I had done, and they came right over and took me to the hospital emergency room, and then I was sent to poison control after explaining to them what had happened. The doctor said they needed to run some tests on me, which might take awhile. It was getting late. The results came back negative; I didn't have a toxic level of poison in my bloodstream. I said to my parents, "I'll be all right."

The doctor pulled me aside and asked what drove me to eat poison—he asked me what was wrong. I explained to the doctor that I'd had enough of living the hard life, and I was tired. The doctor pulled my parents aside and said he saw these kinds of situations all the time. He advised them not to sit on it, and to take me over to the psychiatric hospital on Arbor Way. They specialized in helping people to deal with their problems, and also they could help me bounce back. That same night, the doctor called right over and set up my appointment.

I can't seem to remember how I got there—either the ambulance took me or my parents did…I'm not sure; I don't remember that part. Finally I arrived at the hospital. My parents were told that the hospital would keep me there till the next day or till I was ready to go home. They said no one could see me. My parents had to leave right away. I couldn't have any outside contact—that was why I didn't remember whether my parents brought me over there. They wanted to get to the root of my problems. I had many issues, but I looked around and realized I was not by myself—many women were there. People were walking around in a daze, like they didn't know what day it was...they were really spaced out. The women who were in that facility had many problems, so it made me feel I wasn't all alone in this; I wasn't the only one going through these kinds of issues...I was not the first nor the last.

I got some sleep that night, and the next day I convinced them that I was just fine; I just needed to go home to my daughters and family. I told them I sat up until I fell asleep, thinking about what could have happened to me and my daughters. They gave me good advice. They said, "It's up to you if you want to continue seeking help—you have to want help; it's your choice." I explained my faith in God, and that I belonged to a church. They felt some relief that I had some kind of resource as far as which direction to go in. They felt comfortable enough to let me go home. They called up my parents, who came and got me. Adora had been watching Kyva for me—she said she'd keep her for a couple of days and let me rest. That was kind of her.

I told my parents I needed to go home—that I would be all right; I just needed to go rest, and I wanted to be alone. I needed to trust my own self again. I got dropped off at home, and went to sleep. I woke up around 6:30 p.m. Yes, I seemed to spend time in the living room. I started talking to God again about my life—I was weak, but I had a grip on myself this time. I told God that I wanted a better life for me and the girls. I told God that I had slipped into a deep hole, and it felt like no one could hear me crying out for help. I started asking God to forgive me and help me—I wanted to live and not die, spiritually and naturally. I could hear God saying clearly, "I never left you, I was always there for you. You see, you're still here. It wasn't your time to go. Your experience is only to help someone else, maybe worse off than yourself." He said, "You don't have to do this alone."

Right then and there, I knew exactly what God was telling me to do. I called two of the mothers of the church and an elder, along with my mother, to come over to the house. Well, my mother gathered them up for me; they came right over and started praying for me. They did not stop until I felt a breakthrough. I can remember one of the mothers walking through my apartment praying—she found something in the back of my apartment in the kitchen area. To this day, I don't know what it was—all I know is she threw it out the door in the trash. Did I go looking for it? No way! I was sitting on the floor in the living room, and they kept praying over me. Finally I felt a burden being lifted over my head.

14. The Power of Prayer

I felt so light, it was like floating in the air. God would always have a ram in the bush, and He'd already had a backup plan for me. I hadn't felt that light in a very long time. Of course I felt my life was seeing a light at the end of the tunnel. I said, "Maybe this is it—my life will finally come together." I started thinking about going through so many doors in my life. I was so drained over my relationship with Michael, and I allowed that situation to take control of me, on top of everything else. I'd always had an opportunity to start other relationships, but it wouldn't stick; that was just to get over the previous relationship. It wasn't fair to the other person, because it wouldn't be real. I said to myself, "I don't want to go around hurting other men because of my past...that would be breaking their hearts the way mine was broken...I can't do that." Not only that, I knew I had to be careful of my choices and seek God first. You don't have to find them—let them find you, ladies, and make sure the relationship is what God wants for your life.

I knew I had to be careful, because I was still vulnerable, and still somewhat broken-hearted. As time went on I kept repeating to myself, "Don't do anything you'll live to regret." I also told myself, "Don't do anything on the rebound." As time went on, month after month, much later I tried to start a relationship, but it just didn't stick—also, he knew my heart wasn't in it, and there was a blockage there. I didn't want to rush myself, but I knew that I needed to get myself together. The one good thing about my life was that a close friend would always tell me to be careful. There were people in the church saying awful things about me—it was always coming from ***WOMEN***. What hurt me the most was that I knew these people very well…I had dealt with these people all my life, and some I had just met, but the things that were being said about me tore me up inside. I didn't let it make me retreat into my shell; they just loved to gossip. Most of it was false—if people would learn to stop trying to control someone else's life and learn to take inventory and more control of their own, maybe they wouldn't have time to be in mine.

Women seemed always to compliment me and smile in my face, but I was told to watch those who were constantly complimenting me all the time; they would chew me up and spit me right out across the table. People would never say anything to me directly, because I was too outspoken and would tell someone off in a heartbeat! I learned how to slow down a lot, but I still tell people, "Please don't cross me the wrong way—every now and then it comes out." I began to put up walls around me when it came to women. I started hating women for a very long time—and I'm a woman, too! They always made their own assumptions about me, but they would never come to me on their own or confront me about anything —maybe it was just pure jealousy? Maybe because of all that I've gone through I always bounce back…that's a sign of strength, or of a powerful woman, and maybe some of them get intimidated. They just assume things about me without knowing me personally.

Some of them spread a rumor that I had AIDS, because I'd lost a lot of weight. They also called me a slut. I couldn't help it if guys approached me during my hard times—but the guys who approached me all had great respect for me, which taught me that not all men are the same; it's not fair to shut them all out due to one negative experience—that's not fair to the right person. I still had balance in my life, going to church and reading my Bible; the scripture said, "Thou shall love thy neighbor (Matthew 22:39 KJV)." How can you say you love God and hate your brother or sister? (I John 4:20 KJV) That's what I heard in church all the time, but I believe God understood my heart—that's why He's God and not man. Thank God, not man!

15. Who Broke into My Rented Room?

I knew I could not get on my feet living the same way I had been living. I had known Shada for years—she loved my daughter, and I stayed with her for a short period of time. She helped me out in any way she could, but she was going through issues herself. She seemed to have it all together—not like some women, who wear it, and you can really tell what they are going through! I found out she was cheating on her husband. Because I was so damaged inside toward my broken marriage, it left a bitter taste in my mouth. I found it to be disgusting—that's how I felt. I believe a relationship should be 100% trust; if you're going to go as far as cheating, leave the faithful spouse alone, and do what you want. I felt guilty knowing what she was doing, and I had to leave her alone—I couldn't take it, because I started feeling angry towards her, and I lost respect for her. What she was doing was her business, but when I see something that I want no part of, I back away from it.

I decided to rent a room for me and Kyva. I thought that it would be better if I had my own space and privacy. I found a place rented by an elderly couple—I believe he was in his 80s. It seemed to be a very nice quiet street, not that bad. We moved in and got settled, but the landlord seemed strange and scary—very scary, now that I think about it. But the place was very clean, which I liked. The second week I was there, I came home late that afternoon and I noticed that the door to my room was open—I saw that someone had broken into my room. I didn't have much, but they took everything my daughter and I had, down to our underclothes. They cleaned us out completely!

I went dashing downstairs to confront the landlord. The only reason I went dashing at him was because he was always home, and saw everything. He knew when I was in and out; even if I was creeping up the stairs, he would not leave his home at all. I started feeling suspicious of him anyway— like I said, he seemed to be scary. It didn't sit well with me; I didn't believe he would tell me the truth, and we got nowhere. I looked at him and said, "I'm out of here—don't worry about the rent I just paid…keep it. I'm leaving." And that same night, I got out of there.

16. Moving in with My Sister Theresa and Her Family

I called up my sister Theresa told her what had happened. She would always say, "Come stay with me."

I said, "You have a family—I can't do that."

I never believe you should move in with a married couple, anyway, but that's just me. She loved my daughter Kyva, and Kyva loved her—to this day she still talks about her auntie! Well, she convinced me to stay with her. Not only have that, she said, "stay as long as you want to." My sister would give me whatever she had—she didn't care.

I did hesitate to stay with her, but finally I said, "What the heck—where am I going to go now?"

I would always pack our clothes in big green garbage bags…to be honest, I felt like a homeless person, living out of bags, dragging my younger daughter along. When I got there she welcomed me in—my niece was so happy to have me there. We were always very close. My sister said I could stay in her son's room. I asked where he would sleep, and she said she had worked that out already, and I should just focus on taking care of myself and Kyva. What a blessing that her son gave up his room up for me and my daughter, and did not complain even once the whole time I was there. Theresa had two kids and was married, but she made sacrifices for me—that meant a lot, and I will never forget that.

I told her that I would help her out in any way I could, but I knew my stay would be very short there, because she had her own family—not only that, she was my sister, looking out for me in hard times. We've always had that special bond. She could talk to me about almost anything. I never wanted to come between my sister and her husband; I was taught never to get in the middle of someone else's marriage...that's what I was taught! That's between husband and wife only! Therefore, I had to get on my feet quickly—I just did not want to cause any problem or inconvenience her in any way.

One day my sister sat me down to go over finances. It was my idea—she did not want anything from me; all she wanted me to do was save my money and get on my feet. I told her what I could afford, which was not bad. She said, "That's too much—pay less." We finally came to an agreement of payment, just the two of us. Also, she offered to babysit Kyva—that was included as well. I tried my best to stay out of her husband's sight; that was just the way I was.

One day I came home from work. It was payday, and I handed my sister some money. The next thing I knew, as I was walking toward my room, her husband came storming at me, asking, "What did you give her?" I did not answer him, but my sister did. He pushed me against the wall and said, "Why are you looking at me that way? Can't you pay more money?"

I said, "NO!"

I looked at my sister—she was so scared of what could happen. She snatched at him and said, "Leave her alone!" I was about to charge after him, but she looked at me like *Please don't.*

I said I was leaving. I called up my girlfriend Poma*—I was lucky to have some good friends in my life. She said, "Don't say a word to him—leave in peace. Throw your clothes in a bag—I'm on my way to come get you now." She told me to wait outside. I don't remember if I called my father or not—I hope not;

he would not have been happy about that. I didn't understand why Theresa's husband wanted more money. She said her rent wasn't much, so what was the deal? I never asked why; I just knew that was my cue to go. I hated to leave my sister, but I felt like her son needed his own room back, and I was taking up space, so I should leave now! I saw the look on my sister's face, like she wanted to cry—it hurt her so bad. My eyes said to her, *You can call me or talk to me anytime—this will not change anything between us*. I realized that God always sends people my way to help me out. ***Thank you Theresa Marie Pinckney***

17. Grateful

I thank God my daughter and I never had to live in the street. Ever since that time, I never look down on homeless people when they come ask me for money. Even if you're not sure what they would do with the money, put yourself in their shoes—they have nothing, and it took something to drive them to that breaking point. I understand now how that is. Finally my friend Poma picked me up. She went to my church, and Kyva would always get up in service and go sit on her lap in church. Poma bonded with her, and became her godmother. From that point on she bonded with me, as well. She would always give me spiritual advice and I listened to her. She made me realize why I was going through some of the things I was experiencing, and it made a lot of sense, even though there were times when I did not want to hear it.

I spoke with my parents about what happened to me at my sister's house. My father said my Aunt Betty lived alone and she had an extra bedroom on the first floor below my parents. I ended up moving in with Aunt Betty, back where I started from. Poma helped me move in. We shared the things I was dealing with. She knew I was having some difficult times in my life. We talked and talked and talked. She started taking me out to eat, buying new clothes for me and my daughter Kyva—she helped me in so many ways. She helped me get my self-esteem back; she made me feel important, like I was somebody, and never looked down on me…not once. We became very close friends. She would always give me a word of encouragement, or pull out her Bible and read me a scripture, at any time, every day and night. She never left my sight until she went home or to work. It was like she was on an assignment from God to help me get on my feet and never look back.

People had a lot of negative things to say about her—they said she was crazy, and all kind of nonsense, trying to tell me different about her, but they never approached her directly. It was people in the church telling me this—it turned me off so bad. I said to myself, "I feel good around her—they are not helping me out in any way, because she's doing the right thing and being a good Christian, helping out her brother or sister. She is always pleasant, and she's a giver, too—what the heck is so wrong with that?" But those people made me feel sour about their behavior toward her. She did not do anything to them but speak her mind and the Word—was it too much for them? I wasn't sure what the problem was, but I stuck with her. To this day, she's in my life and they are not—she still checks on me, and they don't. I'm trying to get over being bitter toward people; stuff like that makes you want to go there and not get out of being bitter, but she helped me get through the things I was dealing with. Thank you, Poma.

18. Doors

I lived with my Aunt Betty. This was nothing new to me, because I spent a lot of time with her during my pregnancy with my first child. My Aunt Betty and I used to sit up at night laughing and talking—she would always make me breakfast and dinner. She welcomed me in with open arms. I was very open with her. She had five kids. One of her sons was moving back home, and the room I was staying in was his room. I believe I stayed with her for a short time, too—I don't remember exactly, but I knew I had to move again. He needed somewhere to live—he was her son and he came first. That was not a problem for me; I've moved around like running water from a faucet. I sat her down and told her I was moving out. She understood. We had a wonderful relationship. She said, "I will be checking on you, okay?"

I said, "Let me make some calls and let you know where I'm going."

I called my good friend Cheryl. She was living with her mother, Christine at the time, but her mom had a good-sized apartment. Cheryl was pregnant at the time with her daughter. Cheryl said, "I know you don't want to hear this again and again—I feel bad that you're moving too much. Just live here and stay put for a while; we'll have to share the same room, but we can bunk up here—don't worry about a thing. I already talked to Mommy; she would love to get up and fix you and Kyva breakfast before you go to work." ***Thank You Christine Hawkins.***

My friend Poma came and picked me up, and took me and Kyva to Cheryl's mom's house. I was so excited to live with them—it was home for Kyva, because she was always with Cheryl and her family anyway. I got spoiled living with Christine; she would wake me up, bring me breakfast in my room and say, "Saundra, you need to eat!" but I would get up and take my food in the kitchen. She spoiled me and Kyva. Cheryl and I would sit up every night and chat sometimes until we fell asleep.

One night we were talking and all of a sudden Cheryl jumped on the bed and started screaming—I had no idea what the problem was. It scared me, so I jumped on her bed shouting, "What?! What?!"

Cheryl said, "I saw a mouse!"

I said, "Big or small?"

She said, "Small."

I said, "Girl, you are 8-1/2 months pregnant, just about 9 months pregnant!"

She had a few weeks to go. She said, "The mouse was tiny."

I said, "Get down before you go into labor, girl!"

We laughed all night long. She started itching at the same time. I said, "Girl, you are driving me crazy!" Then I started itching! We fell asleep, and the following week she felt pain. I started thinking she was about to have her baby. I did tell her I had to give her space, and I'd find me somewhere to live for sure; she shouldn't worry about me. But she didn't care—if I'd stayed with her, she would have worked it out, but I did not want to put her in that bind. I started preparing myself to move again. I knew I didn't have enough money to live on my own yet; not only that, but I still had to provide for two daughters and daycare.

So I applied for Section 8 affordable housing, but they

turned me down, saying I made too much money. Can you believe it? I mentioned it to my friend Poma , saying I didn't want to bring my daughter to a shelter, nor did I want to live in the streets. Poma was married and she had her father-in-law too—I couldn't live with her. My older daughter lived with my parents, so I didn't expect anything from them—that was a lot to ask of them, so I did not bother them.

Robin A. King, girlfriend

19. Helping a Sister Out

Another dear friend of mine I met from church was Robin. I told her my friend Cheryl was about to have a baby and my daughter and I needed somewhere to live. She had recently just moved back from California, and she had kids—I felt bad asking her, but I had to be with people I was comfortable around. Not only that, I had to trust them around my daughter, and it had to be someone who would allow me to bring my daughter Anisia on weekends. That was not a problem for Robin. Not only that—she was Kyva's godmother, too!

Robin and I were friends for years—we worked together, sang together, cried together, and she was my spiritual advisor. I was also the godmother of her baby daughter—we were like family. We were always close and always kept in touch, even when she moved to California. She made sure I stayed up on my singing career. She said to me, "Look, you need to apply for some type of housing assistance—there's nothing wrong with getting other help. Girl, you've been through too much, and now it's time to go get some help."

Robin said I could live with her; it was not a problem. "You and Kyva are welcome to stay with me," she said.

I moved in with Robin. Poma helped me move in. I was working the overnight shift, and so Robin said she'd keep Kyva while I was at work. She put her kids to bed on time, and so I knew Kyva would have to do the same. She helped out in any way she could. Letting me stay there was enough—she also gave me wise advice, and steered me in the right path, which was more than generous. I didn't even have to go clothes shopping—she was one sharp-dressing woman with class. One day she was cleaning out her closet, which happened right on time; she gave me clothes, handbags, shoes—name it I had it. Well, it was a brand-new wardrobe for me…that was my girl Robin.

Robin asked me how many times I had moved. I said, "So far, eight times. I am tired of moving around…it's so draining. No regrets, because you all helped me out, and I'm grateful."

She said, "God is with you every step of the way—no matter what, don't lose your focus in Him."

There were times I felt that God didn't hear my cry—I felt I failed Him, and therefore He did not want to hear me, but that is not true. **Matthew 28:20 says, "*He said lo, I'm with you always even till the end.*" 1 Peter 5:7 says that He will never put more on you than you can bear: "*Cast all your care upon him: for he cares for you.*"** Well, things started to get better. Because of so much that has happened in my life, I always started expecting something bad to happen when everything was going right for me—that's the way I thought. I felt that I was born for hard times. I stayed with Robin for a few months, but I knew what I had to do—I still had a good job working at the Le Meridien Hotel in downtown Boston. I was working the overnight shift, and my salary wasn't too bad—I loved that job, I got good meals, and everybody there looked out for me—they said I was the girl who "didn't take no stuff"! That's just I how was, but I was very nice to everyone there. I would get a ride to and from work, and the hotel chef would always cook me up something—I had it good there.

My supervisor's name was Karrissa. She and I bonded, and we started sharing each other's story. She said, "You need to ***write a book one day.***" The next day I started writing about my life—you see, I started this book in the early '90s and have continued up until now she said, "I don't know how you kept your right mind going through what you've been through—it's God. If you ever need my help in any way, let me know what I can do is push for a raise, and get you on the day shift, though that may take awhile."

I said, "Never mind—I'll stay on nights. It's more money, and if I get a raise, that's good for me."

Karrissa asked if I had applied for Boston Housing Authority. I said I hadn't; but that I had applied for Section 8 and was on the waiting list. She said I should go apply for Boston Housing Authority…it might take awhile, but at least I could get started on it. So I did just that.

20. The Last Move!

Karrissa called me up and said, "Let's talk—tomorrow come by my house." The next day I went by Karissa's house on my way to work. She said, "Look, I live alone—it's very nice for your friend to let you live with her, but you and your daughter should come live with me until you hear from Boston Housing Authority. You don't want to keep hearing this again, but I feel this is your last move until you get your own place. Saundra, please come stay with me—I'm never home on weekends; I travel a lot, so this would be like your own apartment, and I don't want a dime from you. Keep your money; I'll talk to the boss lady about getting a raise. Even better, I'll watch Kyva at night so you won't have to bring her all over town—just leave her with me. She's in good hands with me, trust me. I have a parrot and she talks to people...don't freak out, but your daughter will love her, You'll be closer to work and you won't have to travel as far."

Well, it all made sense—I thought about it for a week and told my friend Robin; she was fine with it. Karrissa was so glad that I took her up on her offer. She asked me, "Mentally, are you okay?"

I said, "Yes, but people think I'm crazy—they said crazy things about me, and they even said I was pretending—that's sick."

She said, "No—because they didn't help you; that's guilt."

"Wow," I said, "you're right, girl! I believe it is guilt... but who cares what they say or think anymore? Forget them.

I need to move on and do what's best for me and my girls, right?"

Karrissa looked at me and said, "I feel this is your last move, Saundra. You have to believe that within your heart, God is with you, and He loves you. This storm you're going through right now—you may not understand now, but you will later, and it will touch many lives after you write your book. Not only that, but you should share your story with others about what got you through—don't worry about being embarrassed or others feeling embarrassed; it's not about them. It's to help others, okay? Now move forward from this point on. Other storms will come in your life, but you will handle it lightly, because of what you've already gone through."

That meant a lot, coming from Karrissa. It's called wisdom. Yes, I was happy living with her—it felt like my own apartment. There were a lot of daycares in that area. She lived in Harbor Point when it was just newly built—it was near the water. I would walk outside just for the cool breeze off the fresh water—it was so nice there, and I loved it so much.

Karrisa kept saying, "Something good is going to happen to you—I feel it, Saundra." She said something good would come out of my trials and tribulations. She was just blowing my mind, time after time. Meanwhile, I stayed in touch with all the people who allowed me to live with them. They were happy that things were coming together for me. Finally, I got a raise at my job! I even purchased a brand- new car—my dream car, a Dodge Shadow Convertible. It a was black soft top that lets down. It was a really nice car, and my self-esteem was coming back. I started feeling like somebody. I'd lost a lot of weight, but started picking my weight up again, though not a lot.

I received a letter from Boston Housing Authority, and I asked Karrissa to read it for me, because I was afraid it was bad news. She opened the letter, and tears started coming from her eyes. She looked at me, smiling at the same time. I said no, but she nodded her head and said, "Yes, you got it." I started crying—we grabbed each other and kept crying, saying, and "Thank you! Jesus, thank you!" I told my daughter, but I was not sure if she understood. Kyva was almost three years old. I said to Karrissa, "Thank you for hanging in there with me, girl."

She said, "You got it—that's what friends are for. I told you God was always on your side, and one day you will share what you have gone through and help someone else."

I said, "Girl, you're right, One day I will tell my story about all the doors I've gone through—but this still may not be the last door, right? I want to be there for someone else, like you and others have been there for me. I want to give back what others have given to me, and more."

The following week Karissa and I went house shopping—we were going crazy buying everything. I purchased new furniture—wow, we had a blast! Boston Housing Authority called me to pick up my keys—I was beyond Cloud Nine; I was on Cloud Twenty! I was done, I was beside myself. My parents were so happy for me and my father said he'd help me move. I was getting a two-bedroom apartment, therefore both of my girls were going to be together. My mother wondered if it would be better for Anisia to stay with them until school was out for the summer, rather than interrupting her school year, since my parents' house was located so conveniently to the bus stop. She could come home on the weekends, like she had been doing. I was fine with that—I didn't want to mess up the system she had.

I moved the following month. Karissa was right there to help me out. She spent her own money, and bought so much stuff—my apartment was fully furnished by the time we got through with it. I had a new apartment, new car, and new life— how about that? I was still singing, and doing what I do best. Yes, it all came together. I moved to Burke Street in south Boston. My sister Theresa lived around the corner. It was a prejudiced area, but she never had a problem, and neither did I when I lived with her at that time. I felt good about myself, buying my dream car and having my own apartment again—not only that, but my two daughters could be together.

I felt that my life was coming together. I was able to put Kyva in the Wee Folks Daycare in South Boston, only five minutes away from my apartment. She was way too active to stay home, and to this day she's never home, always staying busy. Everything was going great, but I was always a little nervous when things were going to good to be true. I was single, and I wanted to stay that way for a while. I wasn't looking at all. I wanted to put my focus on my singing career. I was always in the church choir, singing with other groups, and as a soloist at weddings, banquets, funerals, etc. My hat fashion was one of my biggest concerns. Not only that—back in the day I wrote a song called ***I appreciate you, dedicated to a dear friend*** during my hard times. So I wanted to finish what I got started. My ex-husband told me that I was going to marry a musician, because he wasn't the one, but at that time I did not want to hear that.

21. How I Met David

Music was my world. I had attended the New Life Restoration Temple church in Dorchester, with Pastor Bishop Bernard N. Bragg. During that period of time, we would get many Berklee College students there who would attend our services. We were told there was a young man, a freshman who had just started school, a musician looking for a church to play at. He started coming to our church. His name was David. He lived near my apartment complex, and my pastor asked if I could give him a ride home. He would always make sure people had rides home; therefore I did not mind taking him home, along with other musicians. David would always offer gas money—even if he didn't have it, he would say thank you twenty times! Still, he would always ask if I had enough gas, and he would give me what he had.

As time went on, David started venting his frustration about life in general, or whatever was on his mind. He was always trying to make a decision about something. Every time I would take David home, we would sit in the car and talk for a while. I would keep my radio station on—I noticed that Anita Baker was always playing, and I came to find out she was his favorite female artist. Most of the songs were relevant to our lives. One Sunday I took him home after morning service, and when I got to his house, he said he was not ready to go home, because we were deep in conversation. We had service that afternoon, and he said, "Let me go home and change—is it all right if I come back for dinner and go to service with you?" I said that was not a problem.

Well, he came by and we started talking. He needed to vent, and I felt like his counselor. Yes, I was a lot older than he was, and more experienced, so I had a lot to share with him. From that point on, David and I started developing a friendly, close relationship. At the time, we both were in the process of making decisions about other opportunities, but our interest was leaning toward each other. We didn't go a day without talking on the phone. We finally made a commitment in a relationship to each other. Everyone thought I was dating someone else—once David and I made it clear that we were dating, **all hell broke loose!** Everything hit the fan, and some people wanted David with someone else as long as he wasn't with me—girls were after him left and right, **but I was the one he chose to be with.**

22. Harassment Began

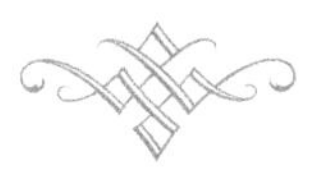

During that time we were dating, my neighborhood was getting a little too prejudiced, and things started happening. It started with my car— yes, my black Dodge Shadow convertible. One morning I got up and went to my car, and I saw that my soft top was sliced. The following week, someone sliced it again—it was really bad that time. I did not have enough money to repair it right away, so I got black duct tape to cover it. One morning I was on my way out the door, and someone put a dead crow in front of my front door. What finally did it for me was that I was cooking and Anisia was standing in front of the kitchen window—and here came a big rock through the window. Someone threw it toward my window, and it had just missed her.

Finally I went to Boston Housing to put in a complaint. They asked me to fill out a report, and said they would look into it. You know how that goes. I confided in David about everything that was going on—he was shocked, and said, "No matter what, I'm there for you and the girls; let me know what I can do." That said a whole lot about him. After I filed the complaint with Boston Housing Authority, things seemed to calm down a bit, but a couple of months later someone was throwing water balloons at my front door, and every time I went out there, no one was there—I remember looking across the street, and there was a school directly across the street from my apartment. I looked and there was a security guard standing in the hallway of the school; I saw his shadow. He acted like he hadn't seen a thing. It was racial; I didn't deserve that kind of harassment, and neither did my daughters.

I went inside the house and called David. He said, "Get in your car and come to my apartment." I did just that. I told him I was so disgusted about everything that was going on, but I'd file another complaint tomorrow.

Well, the next day there was dog graffiti all on my front door. I reported to management office again. I said, "Something has to be done—I'm sick of it! What are you going to do about this crap?" I went home and called the police, and filled out another police report. I told the police about all the things I had to deal with living there. They hired a private detective to take on the case, for investigation, therefore he was on it 24/7. The detective called me, and said he'd come by every single day to make sure my daughter and I were safe. I felt safe with him. Later on in the month, everything got very quiet for a while. I was working the overnights, and David said he would look after the girls whenever I couldn't get a babysitter at night. I was still working at the Le Meridien Hotel. Everything started going smoothly again ever since that detective got involved.

I remember one night I had to go to work, and I asked David if he could look after the girls for me. He said yes, he'd stay until I got off work. As I was getting ready, David came by—I asked him if he could drop me off to work, so I could leave the car with him just in case of an emergency. I lived ten minutes by car from my job, and my daughters were asleep, so I asked David to drop me off at work, because traffic was very light. When David came back, he called me at work and said someone broke into my apartment—the girls were fine, and they were still asleep. I believe my older daughter coughed, and that may have scared them off. Thank God my daughters were safe. Minor things were taken, but I'd had enough.

I filed another complaint at the management office; I told them I want a transfer out ASAP! Also, I filed a complaint against them for letting me go through all that humiliation without doing anything about it. I then applied for Section 8 but still got turned down—they said my income was too high, and I making too much money. I couldn't believe it.

23. Putting in Another Housing Transfer Request

Finally, because I was making so much fuss, they did finally transfer me out. I was transferred to Heath Street housing complex. For a while I loved it, until the drugs started to get heavy in the area. We would come home and there would be guys in the hallways, but they would never say anything to us but hello—they were very nice to us, but I didn't feel comfortable with the girls there. The hallway would always get flooded out every time it rained, and it was dirty. I would call management every week to come clean the hallway—sometimes twice a week. I was going out my mind. I said, "This can't be happening—can't there ever be a happy ending to the story?" There was getting to be too much traffic in my hallway, and I knew it was time to go. I put in another transfer request with Housing. Finally I got accepted into a nice townhouse; it didn't take long at all. Meanwhile, David hung in there with me the whole time—we were talking about marriage. We knew we couldn't be together like we wanted unless we were totally committed to each other. We both agreed that we wanted to get married.

24. More Hell to Pay—1996-1997! Door

Ever since everyone found out how serious David and I were, and when on top of that the rumor got out that we were planning on getting married, it seemed like I had trouble from everyone in the church and out of church—but mostly in the church; that's where I spent most of my time. They were so against us being together—what a shame! A couple of close friends of ours would always pray with us, and encourage the two of us to stay together no matter what people were saying and plotting behind our backs. Some of the plotters—we knew who they were. My dear friend Poma didn't put up with anything; she tried her best to let no one get in our way, but we knew we had to hold our own. We had very close mentors, Ann & Joseph Davis Sr., who would always give us wise advice; they told us to keep our business to ourselves, and also they would take us out to eat, and mentor us. David had a very close relationship with his mom; they shared everything. I guess she felt I wasn't the one for her son. So many people got in our business, to the point I thought they were obsessed over David—it was very bad. I thought some of the people at church had my back, but during that time, I found out they didn't. Some of the guys wanted me, and some of the girls wanted David, even out of state, or other people wanted to play matchmaker for David. People started calling down to his hometown, and making false accusations against me…rumors kept going back and forth. Even though we went through so much hell just to be together, we said we were still getting married—who cared what other people thought?

We planned our wedding. He called his close friends and family members, and I did the same. Eventually, he told me that no one from his family would be attending. I said, "There's nothing we can do about that—do you still want to go on with this?" With tears in his eyes, he said yes. Here came our wedding day—it was set for 2:30 on a Saturday afternoon. Around 2:00, half an hour before the wedding, I received a call from David. I saw his number on the caller ID, and when I answered it, I thought it was weird that he was calling me.

David said, "Saundra…." Then he hesitated. He said, "Saundra…" again.

I said, "David, what?" He paused. I screamed and shouted, "I know this is not what I'm thinking—no no!"

He said, "I'm so sorry—I just got off the phone with my mom. She said if I marry you, she's going to commit suicide." Because of David's loyalty to his mother, he did what he thought was best to keep the peace.

I said to David, "I can't believe you let her do this to you!"

I slammed the phone down, ran out of the room, and lost it. Meanwhile people were waiting at the church, the limo was outside ready to go, my mother had worked all night and morning to prepare the reception dinner, everyone in the family was dressed, some were still getting dressed—it was a freaking nightmare in my house. I asked my friend Poma if she could go to the church and announce that the wedding had been called off. I was trembling, crying, and confused about men—not only that, but my daughters had to watch me go through that, and then I had to explain to them what had happened, in my own way.

I really thought I was dreaming! It wasn't a dream, it was

reality, but it was the kind of stuff you see on TV. I hated what he did to me; I hated the way I felt. His mom didn't want to accept the fact that her son loved me. She never got to know me for herself—all she saw was that I had two kids, and I was ten years older than David. People were judging me in all the wrong ways. David's mother and I would have conversations on the phone, and it was not pleasant at all. I was not going to allow someone talk to me like I wasn't human. This was all because of the love she had for her son—that's fine, but he was ready to make his own choices in life. You can't control your kid's life once they are of age, and on their own. People were still calling in and out of town, just putting me out there—worst part was that these were people in the church. I've learned in life never to judge somebody as if you are perfect and can do no wrong—who should cast the first stone? People were saying I looked very old. The situation was so bad, and I'm shocked that we still held on to each other. Now you see why the wedding was called off—he had to make a decision between two women, which wasn't fair. The Bible says, in Mark 10:7, **"When you become one in flesh a man leaves his father and mother, and cleaves to his wife."** I was not asking him to break off his relationship with his mom, but I wanted him to realize he was a grown man, and needed to make his own choices in life—if not, he would never be happy, but would always be paying attention to someone else's opinion. A couple's marriage should be their first ministry, and everything else should be secondary.

Me & Sister Wilma Faye

25. Comfort

Sometimes you're so full of hurt that it feels like there's nowhere to pour your hurt out. The very same day the wedding was called off, one of my sisters, Wilma Faye, came to me, and said, "Give him another chance. Maybe it's not the time right now—timing is everything, and if you're meant to be together it will happen no matter what."

I was so hurt that I didn't want to hear that, because I was so bitter inside, but Wilma Faye had powerful words of encouragement. She worked with me until I calmed down. At the time, I felt that I hated everyone who had interfered in our lives. I still didn't understand why people were pulling against us so hard—it was awful. I knew God was not pleased. Later that evening, Wilma went to pick up David, and brought him by the house. He also had a close friend from his hometown who stood by his side. He came by the house that evening; that was the first time I'd seen him since he called off the wedding. I didn't know what to say or how to react toward him—he looked so pitiful. He looked at me with tears in his eyes and said, "I still want to marry you Saundra, but not right now." He kept saying he was sorry. We talked and talked and talked until I fell asleep.

The next day I woke up early. It was Sunday morning. David and I were not sure whether we wanted to go to church, but we did end up going—we both felt so weak and confused about people. It was rough again for me. My pastor called me and David up for prayer. People were surrounding us, and some of them were people I didn't want to be next to. I felt that I was fenced in, but they kept praying and praying. David and I embraced each other and started crying; he looked at me and told me that he loved me, and we would be together no matter what. I realized I had to leave it in God's hands. I freed myself of the way I was feeling, and asked God to help me deal with all the anger I had inside. We both made up our minds that we wouldn't allow anyone else to come between our relationships. **"If God be for us, who can be against us? (Romans 8:32)** God knows all things work together for our good.

As time went on, acceptance started taking place. David and I made sure we weren't going to let anything come between us. Later on in the month we went to the justice of the peace; we felt like we wanted to do it in private—so many people were against us, so the heck with it! On March 14, 1997 we got married. We had very close friends who had our backs, and my side of the family; the next day we both were on Cloud Nine, but I knew it hurt David that his family was not there. We had to do what we had to do. David finally ended up calling his family to share the news. He and I had such a tight bond with each other—David and I never spent a day apart from one another. **"What God hath joined together, let no man put asunder" (Mark 10:9)** What makes people think that they are more powerful than God? David and I had such strong faith; we would always talk to God, and seek his direction. David always wanted the acceptance of our marriage from his family, and we always sought God about it.

26. Acceptance

Things started to come together between me and his mother. I wanted her to know me for herself, and not based on assumptions or what others said…it wasn't all about what other people wanted; it was about what David and I wanted. When you leave it in God's hands, He will work it out for you. He knows how to do things in His own timing; our timing is not God's timing. We all started to come together, after being married. Four and a half years later, David, and I felt that we wanted to have a renewal of vows, complete with families, close friends, and relatives. I asked his dad to walk me down the aisle along with my dad. He said he would love to walk me down the aisle. Kyva loved David; his dad would always make us laugh. We would spend time with David's family, and he would spoil her with chocolate candy! One day we went to visit David's family out of town. David's father got up and took my daughter and his son to the store, and came back with a big bag of chocolate candy, and then gave it to my daughter. Guess what she did? She ate the whole bag! Later that night I knew it was coming—she had the worst stomachache. Gee, I wonder why? We laughed and laughed, and I got her something for her stomach. She wouldn't do that again! His family came up to Boston that same year for Thanksgiving, and my family came over. I cooked dinner and made dessert, and we had a good time. My mother cooked, too. His father ate well.

27. Our 2002 Renewal of Vows

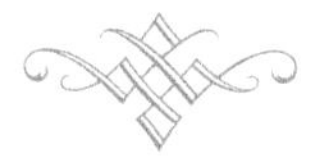

On June 2, 2002 at 2:00 p.m. at the New Life Restoration Temple Church, in Dorchester, Massachusetts with Pastor Bishop Bernard N. Bragg, David and I renewed our wedding vows. We had nineteen bridesmaids and nineteen groomsmen—two maids of honor, two best men, one flower girl, one ring boy, and four hostesses—yes, both families joined together. My pastor performed the ceremony. We had great musicians, and singers, and dancers—**what a great church wedding**! We danced, we cried, and we bonded. All things worked together for our good. Of course, I could not let the wedding end without saying something to David. I asked my pastor to hand me the microphone. He hesitated, but I nodded my head and looked at him saying with my eyes, *Can I have the microphone?* My pastor Bishop Bernard N. Bragg knows me well, but he didn't know what was going to come out of my mouth—he looked like he was saying, *Oh God... what is she about to do?*

I took the microphone, looked at David in the eyes, and started prancing down the aisle singing "I Am Telling You I'm Not Going" (by Dream-Girl). I lost myself in that song—I could not stop belting that song out. Those words meant a lot to me. I put all my gut in that song, so he got the message—you're gonna love me! After that, I gave the mic back to my pastor. Yes, we exchanged our vows again! It was done! My pastor laughed and then took us to church and we danced and danced. What a joyous time we had—two wedding receptions for those who weren't able to attend the big reception. It was held at the Le Meridien Hotel, which was a fancy hotel, very classy—yes, I worked there and was very proud I did. Now you see everything came together after all was said and done. What a great reunion we had! "**God's word shall not return unto me void.**" **(Isaiah 55:11)**

28. Making Wrong and Right Choices in Life

David and I had gone through so much; we wanted to make a fresh start somewhere else. We were always talking about relocating. We would talk to our daughters about it, but it didn't sit too well with my older daughter. But she said, "Do what you have to do." Kyva was ready to go. David always wanted to move back to his hometown, and I wanted to move to Los Angeles. At the time, we wanted to make a change from everything. We kept saying, "Let's pray about it first, don't stop praying, ask God for direction." We decided we wanted to move where we didn't know too many people—we just wanted to blossom somewhere else. We both felt stagnant, not growing, not appreciated—there was so much negativity around us, and we were tired of people around us stabbing us in the back, especially when they said, "We have your back." They might have had his back, but if they didn't have mine, they didn't really have his either, because we were one. We had such a bad experience—we felt that if anything good came our way, we were going to pursue it.

Don't get me wrong—as you can see from earlier in my book, there were some great people in our lives, and they helped in so many ways. They inspired me and David, and that's what helped keep us together. God does put people in your life, some for a season. David said, "If we get an opportunity, I'm out! If it doesn't work out, at least we were brave enough to step out, and try it out for ourselves, and not get stuck."

People make a lot of right and wrong choices in life—that's a part of life and growth. There's nothing wrong with wanting to do something different, but you have to make sure it's the wise choice. When all is said and done, it's all about what God wants for our lives. There is a purpose for everything in life. David was traveling back and forth to Houston, Texas working on a CD project. They fell in love with him. He had good friends there that he had known for a while; he got a very good job offer down there. He was asked to be the music director at a church, and would get paid well! Not only that—we were told that if we relocated there, David would get a good salary, and Pastor Bizmo offered to buy us a house, fly us there, transport all of our furniture and vehicle down there, at no cost to us. Also, he'd help me get a job. Wow that all sounded so good, and we wanted to leave to start fresh. Pastor Bizmo said Kyva would go to one of the best schools there, and it was a nice church. I was thinking about meeting different people—that sounded good to me. David started commuting back and forth between Houston and Boston. He thought it was time for me to take a trip there, so I went down in January. The weather was so warm—it was weird!

I told the pastor, "The only way you're going to move me down here if I get the house of my dreams!"

David and I sat and talked with him and his wife—when I said that, her eyes popped. I said, "No way I'm giving up everything I have in Boston. If I can't get what I want here, there is nothing here for me." But I knew if David moved, I would end up moving too.

29. What Blew My Mind!

Pastor Bizmo said "Let me take y'all around and show you a couple of houses."

David and I were blown away by what we saw—Texas had some of the nicest homes I'd ever seen in my life. There was this one particular house that caught David's and my attention. What a place! This house had a big lake in back of the house, a waterfall that lit up at night in the middle of the lake, big surround sound outside on the patio in back, and when you walked in there was a nice foyer made of all thick glass. It had a nice Jacuzzi and his and hers bathroom sink! There were Roman columns inside the house, marble floors, a humongous kitchen, everything stainless and glass, a built-in stove from the walls, a wine-tasting area, and four bathrooms—every room had its own bathroom and space. You could go to the back patio from each bedroom, an office room, huge laundry room, tall ceilings…I mean, the house was amazing. It was in a very quiet area, with good neighbors, and after that last visit I was done.

I asked David, "What we are going to do, since we both are ready to jet?" We started praying more and more about our decision, and when we got back to Boston, we both felt like we were ready to go back to Houston. David was stalling about his decision; because he was the man of the house, he wanted to do the right thing for our family, but he was not happy living in Boston. That pushed our choice more toward Houston. David received a call from the minister of music at the church in Houston. She said, "What are y'all going to do?"

I said, "It's not me, it's David."

She said, "If he's not going to take the position, we will have to get someone else, but we really want him."

She spoke with David on the phone; after he hung up the phone with her, we sat at the kitchen table. We both agreed that we would be relocating, so then we sat the girls down. I knew Anisia was about to finish high school and would be off to college the coming September, she was graduating June 2005, and she was still back and forth at my mother's house, even though she still lived at home with me. David said they wanted him by February 2005, but he did not want to go without his family. He said we would all go together. He called back and accepted the job offer as musical director at the church in Texas. We spoke with my parents about our decision, and my father said it sounded too good to be true—there were too many cards in one man's hands. We met with our pastor Bishop Bragg in Boston; he gave us his blessings. We told close family, and friends, and the word got out big time...my family ended up giving us a going-away party. I worked at the City of Boston Mayor's Office. I'd been with the city for almost nine years. I gave my notice, and gave up our apartment. I was leaving a lot behind, but it was a new start for us. I knew my older daughter could always come visit during her school vacation. It was like she went off to college, just like other students would go away to college.

The pressure was on. It was getting closer to the time to relocate, and we started packing up all our things. I got rid of a lot of my nice furniture, which I gave to friends and family. We are not talking little stuff—this was big furniture...a china cabinet, my entire kitchen set, serving table, etc. Finally we gave our notice to our landlord; they said we were good tenants, and wished us all the best. Pastor Bizmo made all of our arrangements to be transported to Houston. He said he would take care of it—not a problem. So that was no pressure on us.

As the time started getting closer, David and I were getting job offers in Boston like nobody's business. We started to wonder why these opportunities would present themselves now that we had decided to leave. David started feeling queasy about our decision, but David always hesitated on everything; it didn't matter what it was, so I thought he was getting cold feet, because he was the man of the family and that was a big step to move his family far away from everything and everybody. I was fine with it, but I saw him struggling. His friend from Texas called and asked if we were getting ready. I shared with her what David was going through. She asked to speak with him, and after they got off the phone, David said, "Do you want to do this?" Then he asked Kyva. We both said yes, and he said okay—he told her we would be coming down together. Meanwhile we gave up everything, and packed up everything. I stopped working, my daughter was transferred to the school down there already, and David gave up his job at the church as Minister of Music and all of his gigs in Boston. So what more was there to think about?

30. What Happened at the Airport?

Pastor Bizmo called and said he would take care of our flights to Houston. We did not have to worry about any expenses at all. We arrived at Boston's Logan Airport and went to check in, but the flight attendant said she could not find our names. She looked and looked; then she called her manager to check for us. My sister Wilma Faye got involved. She said, "It looks like y'all don't have reservations." David started feeling nervous. I was getting upset, and said to myself, *Oh God, don't let this be a dead dream that never existed!* My mother, sister, older daughter, and adopted brother, James Hills, were all there. They said, "We are not leaving this airport until you all are all set."

So my sister called Pastor Bizmo and asked him to explain what was going on. He said he had booked the tickets, and she asked where. She gave the phone to the flight attendant, who told him there was nothing in the system. He said, "Give me a moment, I'll call right back." He called a few minutes later, and spoke with the flight attendant. The tickets were booked. David and I did get cold feet, but what do you do when you've given up everything? We boarded our plane, and an hour into our flight, David and I looked at each other and said, "Wow, we feel different already!" It's like we were away from everything that tore us up. We just wanted to start a fresh life all over again. Kyva was in her glory; she was smiling ear to ear. So we said, "We're taking a chance, and if it doesn't work, we'll come back." I didn't want to go back home; it was like I had shut a door that I didn't want to open again. Hurt will do that to you.

So we arrived in Houston, and David's dear friend picked us up from the airport. It was warm and very nice; she took us to get something to eat. We were getting pampered there. She took us to the home of Pastor Bizmo and his wife. We got settled in; our house wasn't going to be ready for a couple of weeks, so we stayed with them temporarily. David and I kept driving to the dream house. We would pull up in front of the house and stare—Kyva kept saying she couldn't wait to move in. Her school bus stop was right in front of the house—how convenient.

The pastor's wife and I became very close. I felt I could trust her, and I could tell she had a lot to teach me. Pastor Bizmo was all about family priorities, so that was up my alley. That's what he would teach David that family is the first priority, even before ministry! I felt that's what David needed—a man who would help him with his family priorities, and put things in perspective as being a man of the house. The pastor was much older than David, so he could learn from him. Pastor Bizmo gave David hours to work down at the church—I believe thirty hours a week. He said he could do all his music there, and not only that, he could work on my CD project. I was all for that. The pastor was really pushing my music out there 100%. He would always tell David, "Your wife's music comes first." It seemed like in Boston David just couldn't tap into my music career; I was never sure why. Pastor Bizmo was pushing it to be born.

I was hearing all of the right things. We joined the church as a family, and they greeted us with love. I fell in love with his members, and his wife—everyone was so very nice to us. Our daughter became part of the dance ministry there. David was the musical director, and I joined the choir. So there we all were, waiting to move into our house. Finally the pastor came to us and said everything was all set to move in, but before that, meanwhile we were still waiting for our transportation, and furniture. I called home to Boston to find out what the problem was. They said they were waiting on a payment; they'd received a check payment up front. They said, "Our guys have been on the road all night long and they need to make that delivery to you soon." So they came to the house with our furniture, the pastor wrote out a check, and a week later they called me and said the check bounced—if I didn't pay it, they would be taking me to court in Boston.

I started to flip out! I spoke with Pastor Bizmo's wife. She called him, and she ended up coming by the house. She said she'd take care of it, and not to worry, it was just a distraction. I went along with it, but I started to have my reservations, asking if this was the right move for us. Later that night after David came home, we started talking. He was complaining about working the thirty hours—he said it wasn't in his contract, and he did not have his freedom to do other things. Finally we ended up moving into our humongous house. It was beautiful—it was like a dream house. We lived near the golf course—every house there was gigantic. You could hear the crickets at night. One night we drove up in front of our driveway, and David saw a black snake! That freaked us out, and of course David felt it was a sign. That night he started praying over me and Kyva. I got up that night—I couldn't sleep.

We started talking. David said, "I don't know, Saundra, about this relocating!"

31. A Door I Didn't Want to Reopen

"It is better to trust in the lord than to put confidence in man." (Psalm 118:8)

David said, "I don't feel right!" But somehow, David and I worked through it. I took off a couple of months before I started applying for a job there. I saw that David was handling a lot of bills, even the ones in Boston, and I wanted to get caught up with the bills, and not fall behind. David was working for the church, and often he would get his paycheck late—sometimes a week late, or even later than that. As time went on, we would say to each other that we loved the house, but we had to maintain the house, and it took money to keep it. Once I started looking for employment, it was so hard for me to get a job there—it was nothing like Boston's job market!

32. What a Shocking Surprise!

One morning our doorbell rang. I asked, "Can I help you?" I was talking through the intercom.

He said, "It's the Texas sheriff." I called David to the door, and the sheriff asked, "Are you Mr. Bizmo?" We gave him our names and he said, "No, this house is in another name."

I said, "You mean this is not our house?"

He said, "No, this house is put up for a lien." David and I were furious! He said, "I know it's not your fault. Speak with Mr. Bizmo, and also we will send him a notice; we'll see what we can do." He was apologetic. This really put a big damper on our lives. David spoke with Pastor Bizmo a couple of times, and said these things couldn't be happening—he couldn't continue to keep getting paid late, putting his family in hardship; there were times when our truck needed repairs, such as new brakes and rotors, and we would have to rely on the pastor, because that was David's income. I got to the point I took out my retirement from the Boston City Hall, just so we would have some kind of extra income. Once I received my retirement check, I had to pay off some bills in order not to go in debt; we still had bills in Boston, and now we had bills in Texas.

33. Rethinking Our Decision

David and I talked about relocating back to Boston. I was told if we did relocate back, it would jeopardize Kyva from graduating from high school, and she might have to repeat her grade, so David and I had to think it over again. We said that we would stick it out, and maybe things would get better for us. I could tell there was a lot of tension at church. People were being let go from their jobs, and it was a small church to begin with, so I said, "Who is left?" That meant more work we had to do. Years ago I came from a small church, and yes, it takes a lot of work—sometimes you have to wear several hats, which can be draining. I said, "Not this all over again!"

David was not happy at all. What pushed me over the edge was that Anisia came down to visit one weekend. When she got ready to go to back to Boston, we stayed with her at the airport to make sure she was checked in. Pastor Bizmo said he would take care of her flight expense coming down, but she went to give her name to the flight attendant, they said she didn't have a ticket to Boston. I went off to the left—I mean, I was going off! We left the airport and went to Pastor Bizmo; we had words. Anisia was so upset that she had to leave the next morning. She felt the tension there. After that, David and I said we needed to find another church to attend; on top of that, we had some very close friends, Trendel and Velvet, who had already planned their vacation with us, way before all this happened. They had planned almost a year in advance to come down. David and I were thinking about calling them to cancel their stay with us—perhaps we could reimburse them what they paid for their tickets? Yeah, right—like we had money to waste like that!

We were still going to the church. They came down, and we had such a great time—David and I had prepared for their visit; we had already gone food shopping. We were so happy to see them come down. We took them to church with us that Sunday morning. It felt like a strong vibe at church. The pastor started preaching—next thing you know, he was preaching on me and David, he just didn't say our names. Our guests could sense that something was wrong…my face looked like *If you say another word, I'm gonna speak out, and it won't be anything nice!* My girlfriend kept tapping me on the back, as if to say, *Don't do it!* David was on the organ. He shut down the music and held his head down the whole time—I could tell he was ticked off, and so was I! But we got through it all.

David and I looked at each other and said, This is it—we are out of here!" We got home with Trendel and Velvet, and we talked about what was going on down there. We said, "Now we trust you not to tell our business back in Boston; we've gone through that already." They gave us their word. Boy, did we spill out our guts—David really poured out what he was feeling inside. It was time for them to leave, but the night before, they prayed with us, hugged us, and gave us their word—your word is your bond! Off they went back to Boston.

The next thing you know—I believe it was two days later—another friend called down from Boston and said, "I can't believe everything you all are going through—I mean some deep stuff. The rumors are out up here, so be careful who you talk to."

I said, "If I tell you the names, will you tell me the truth?"

They hesitated and said, "Yes, I'll tell them I told you." That really made me think they were telling me the truth. I mentioned the names, and they gave me one of the names; I said I did not hear about the other name. At that point I let David handle it, because I would have handled it way differently than he did…we were broken again! It was so painful. David and I called our pastor back at home in Boston, and told him what was said to us—also we released the name of the person that told us. They were willing to face the truth about what they had heard about us—that took a lot of courage, to stick their neck out for us. I went into our room and started talking to God about it. I said, "Why are some people always putting people's business out in the street? Why! This has to stop!"

I started talking to God, and He said to me, "This battle is not yours." The scripture that came to my mind was "**Cast your entire cares upon him, for he cares for you.**" **(I Peter** 5:7) God told me that if I would cast my cares upon Him, He would make my burdens light. My mind was like a ticking time bomb—not only did I know that David was hurt, as the man of the house, it was like tearing down his manhood…it was an awful feeling, and hurtful. We were back to not knowing whom we could trust. David handled it, and told me to let it go. That was hard, but I finally did. However, David and I held bitterness our hearts, and that was wrong! It's like our friendship went down the tubes.

The Bible says, "**Let all bitterness, and wrath, and anger, and clamor, and evil speaking, be put away from you, with all malice." (Ephesians 4:31)** David and I sat down, not only dealing with that, but finding another church to attend, and also a place to live. I said, "What are we going to do?"

He said, "We can't move back to Boston, because of Kyva, so we have to stick it out till she graduates. We have to step out on faith; therefore I'm leaving the church, and looking for another house for us to live in."

We both ended up talking to Pastor Bizmo—yes, I spoke my mind, David said what he had to say, and that was the end of that chapter. We told the pastor that we were moving out the following week. We didn't bother to tell him where we were moving—well, we didn't have a place just yet. We started going house shopping. It was draining, but we didn't have to look too far. Finally David looked at a nice house—it had another lake across the street, a gym, a very quiet street, and the house was newly built! We spoke with the owner of the complex; he brought us into the office. For some strange reason, we told him what had brought us to Texas, and what we had gone through.

David started bonding with him. He said, "I'll help you get a house, but it has to be on David's credit score," because he ran both scores, and David's was approved credit with no money down—only $1000. That was it! Talk about somebody jumping for joy! Meanwhile, he said, we had to wait for all the paperwork, and minor things they had to do to the house, so he put us in contact with the realtor, and consultant, and they said they'd put us up in a hotel temporarily, until the house was finished. That was during the time of Hurricane Katrina, so most of the hotels were full. We were living off my retirement money, but it was running out. The owner of the complex offered to help us move out.

He was floored when he saw our old house. He said, "How could you afford something like this? Oops, you told me already! But wow, man, this is very nice."

We moved out and they found a hotel for us—it was nothing nice. It smelled like cigarettes and old rugs, bugs were crawling all over the place, people were standing and yelling outside of the patio—it was awful. I went to management and complained. They said there was nothing they could do, and

I'd have to speak with upper management the next day. Kyva wanted to stay in the room; she was very tired and did not care, but I'm very picky, and I was not sleeping in that room at all. We slept in the car, got up the next morning, took a shower, and out of there, called our consultant, who hooked us up in another hotel, which was much better.

My money ran out, but David spoke with a close friend of his from Houston. He offered David a job at another church. David took it, but we knew that we had a lot of catching up to do; we needed to borrow some money, but we called people we knew we could trust! We ended up getting money to get us started. David started looking for another job, but for some reason I couldn't get a job no matter how I tried. Finally David got a job interview at Wal-Mart. I said, "I'm going with you on your interview." I went with David, and while I was waiting I put in an application. By the time I wrote down my job experience and handed it to the manger, she said, "Wow, you did all this? Wait right here." She came back and offered me a position, I said, "Thank you, God!"

I said to her, "I'll do anything but cashiering—I don't want to handle money." I drew a big red mark that said "no cashier," and figured I should be able to get what I was asking for.

She said, "Do you want to work in the office?"

I said, "Yes, ma'am!"

"Well, we are a brand-new store that will be opening up soon, and we would love to have you."

David was shocked that I got a job, and so did he—I could not believe that David and I would be working together. David and I started working, we got moved into our new home, and the mortgage was reasonable—things started to look good on our side. But two weeks later, David said he couldn't do it anymore. He said, "I'm stocking cans on the shelf. That is not good for my hands—I'm a musician, and if something falls on my hands or I get injured doing this kind of work, I'm done...I need my hands to play my music."

I understood where he was coming from. David stopped working at Wal-Mart, but I loved it there, and they asked me to sing at the store's grand opening. David brought his keyboard, and we went to town—it was great! I wore many hats at that store. I worked as an office clerk/back door receiving/community outreach event planner. I had my hands full. Later, I put on two fashion shows at the store for promotion—it was great, and I enjoyed doing it. Before I started working, one Christmas we didn't have much money—I would live off the change in David's pocket to buy food for the house. All I had was $10 one Christmas, in 2006. I went to Wal-Mart and was able to buy two cheap watches—one for David and one for Kyva. I cooked dinner, and that was it. We looked at each other and said, "This will be the last time we spend another Christmas like this." The weather was warm, people were walking around in shorts, we had no family down there, and it was horrible! We didn't have enough money to send for Anisa to come down.

I met some very good friends at work. They looked out for me—you can tell when people are in your life just for a season, but they were real friends to me. I adopted some of them like my very own family—there was one friend I could rely on her for almost anything. She bonded with me, and I saw that she was having problems in her marriage, just like all the things I went through. I could see anything coming my way—even if my eyes were closed, I could feel it...that's how alert I was. She would drive me home from work and vent, and we helped each other out a lot. There were two good friends of mine, Teisha and Kendall—I'll never forget them. I found myself being a counselor at Wal-Mart; even with the upper managers...it was because of all my previous experience with issues on top of issues.

As time went on, with all that David and I had gone through, we never thought about ourselves as human beings… maybe we needed to see a counselor; mentally, we were drained. But we kept it going. David got other job offers with his music, called gigging, so he was always busy playing out somewhere. I would go with him some of the time, but because of my work schedule, I would always make sure I got my sleep—I don't play around when it comes to my sleep. I made sure I would have dinner ready, the laundry done, and the house clean. It was another big house, and nice, but I kept it clean at all times, and still worked my full-time job.

34. Kyva's Car Accident

David and I felt that we needed to get rid of our Ford Explorer, because it was giving us problems, and we got tired of putting money into it...we talked about selling it. Well, we did just that. The old man whom we sold the car to was grinning ear to ear—he knew what to do with the truck. It was out of our hands at last—now we were down to one car, a Ford Taurus, which was roomy and nice! But it was hard having only one car, because David would gig a lot; therefore Kyva and I had to stay home. My friend Teisha would take me to work as needed, or David would have to pick me up and drop me off. Not only that—he would have to take Kyva to school sometimes, whenever she missed her bus. It caused us some wear and tear, but we had to do what we had to do for the time being.

Eventually, I was going to purchase another vehicle. Well, one evening Kyva wanted to go to the store. David said, "Okay, but come right back, I may need the car." Off she went. It started getting late, and we were wondering where she was. Later we got a phone call from the Texas police, saying, "Your daughter was in a car wreck! But she is okay."

I started shouting, asking where she was and what happened. He put her on the phone—she was crying like nobody's business, and David was upset! We were trying to figure out how we were going to get to her with no car! We came to find that one of the guys she picked up on her way to the store gave us his mother's phone number, and said to ask her if she could pick us up—he said, "Just tell her you're Kyva's mother and she was in a accident; she knows you don't have any family members down here."

I called his mother, and she came as quickly as she could to pick up David and me. The police pulled me aside and explained what had happened. Apparently my daughter picked up a friend on her way going to the store, because he needed something from the store too. She let him drive to the store. She saw her ex-boyfriend on her way back, dropping her friend back home. He saw that she let her friend drive the car, and when her friend got out, her ex-boyfriend jumped in the driver's seat while Kyva was getting out of the car. Panicked, she jumped back into the passenger's seat, because she was afraid that he'd take off. He sped off and lost control of the car. He did not have a driver's license; he was only fifteen years old. He sped off on two wheels. The car was tilted, doing 90—he slammed into three other cars and landed in front of two more parked in front of their garage door! So there all three cars were totally damaged, and my car was totaled, so we had no car to drive at all. I did not have full collision on my insurance.

What a nightmare. Kyva went to the hospital, but she was okay—just a black eye, and busted-up lip. Thank God the air-bag saved her…it just burnt her lip a little bit. Kyva always told me I'm hostile, and she was afraid to tell me anything, because of the way I would react. Well, I could have torn her ex-boyfriend up when I saw him at the scene. I guess she was so afraid that she told the police about her mom, and when I got there the policeman followed me around, but he understood my frustration. But I proved that I was not going to hurt him. I blamed Kyva for not doing as she was told, or this would never have happened this way. So I didn't jump on the kid—people shouldn't under estimate me.

35. We'd Had Enough of Houston

The next day, my friend Teisha picked me up and took us to the car rental place. David had all kinds of gigs that week; therefore he needed a car, so Teisha offered to take me back and forth to work. I agreed to give her gas money, even though she wouldn't take it. I had to rely on her a lot. The following week, here came Hurricane Ike! The electricity was out for 2 1/2 weeks. We'd had enough of Houston. David and I said, "Enough—we are moving back to Boston."

I called my family to give them the news. They couldn't help me, being that far away, but when we said we were going back home, my youngest sister Barbra Mathis (Necey) said, "Send Kyva back home, and I'll try to get her in another school here in Boston. Why don't you and David work on doing what you need to do, like packing things up—it may take awhile, but let me take care of Kyva so you can focus."

I took her up on her offer. My sister Wilma offered to help out as well to get her back in school, since it was so close to graduation. They were able to help get Kyva in another school. That was her last year in high school, and I was nervous. I went to Kyva's school in Texas and spoke with her counselor—I told her everything we'd gone through with no family there. She understood where we were coming from, and she was able to transfer Kyva's credits to the new school in Boston. I was told by her counselor that she already had enough credits to graduate from high school. I was relieved, and off she went, back to Boston. God made a way out of no way—a longtime friend of the family was able to help Kyva get back into the Boston school with no problems, and she was able to graduate on June 2009. A special Thank You to my sister Barbra! for allowing her to live with you and your daughter for almost a year, and thank you for giving her a sweet sixteen birthday party. I will not forget what all you've done for her.

Kyva's sweet 16th b-day

Youngest sister Barbra & My daughter Kyva

daughter Kyva

36. Returning to My Hometown: Boston

On December 22, 2008 David and I returned to Boston—finally we made it back. I was free at last, thank God. My family was glad to see us back at home, and we were able to be home in enough time for Christmas 12/2008. Not only that—I was home with both of my daughters, and we were able to be back for both graduations the same year, 2009. Anisia was graduating from college on May 9, 2009, and Kyva from high school, on June 8, 2009, so David and I had two big parties to plan. That was good timing coming back home, rather than having one graduation in Texas, and flying back for the other. It worked itself out for all of us.

Anisia Berklee College

Kyva high school
Another course to college

37. Nothing Ever Stays the Same: Things Change, People Change

We had our work cut out for us—that was one of the most important things to me in my entire life, to see both girls graduate. I salute them both! Not only that, I had people looking out for me for employment. During that time, the economy was very bad! But God still had my back; two months later I had a job as an assistant manager working in retail, thanks to Willie Israel, my goddaughter's mother. At the time, since we came back sooner than we'd planned, my sister Wilma Faye called and told us we could stay with her and her son Darnell until we both were able to get back on our feet. We lived with her for about three months. Later, my sister Barbra *moved* out of the first floor of my parents' house, so David and I moved down to the first-floor apartment in April 2009.

I felt now that we had moved and had our space, we could get settled in and put our family values back on track. The only problem I had was that I thought once I moved back home it would feel like it used to—***Who was I kidding?*** Nothing stays the same—things change and people change, and life goes on. That doesn't always mean it's a bad thing all the time. I felt that I had to re-adjust my life all over again—home, church, job, etc. It was very different. Moving to Texas matured me in so many ways. I was not the same, but I'd learned to do things alone. I really learned how to budget my money, and to shop without overspending. I learned how to say no when I don't need something. If I could budget on $5 for the week—well,

I could do almost anything. All of that came from everything I'd encountered in Texas—I had no other choice. The most important thing was that I had to rely on God most of the time. He's the only one who can move my situation around through tough heavy storm days, and sleepless nights. Since I'd gone through rough storms in my life thus far, at this point of time in my life, it seemed like nothing fazed me at all—it was just another wind passing by. Some things just didn't make an impression on me; it couldn't be worse than what I'd already gone through.

38. Never Say Never

However, I'm trying to tell you something—*NEVER SAY NEVER.* Never say, "That won't happen to me," or "If that were me, I wouldn't put up with that!" Maybe you say, "I would have been gone!" We've all said something like this, right? It's like right when you get settled down and things seem to calm down, and it looks like a light at the end of the tunnel, boom! Here comes another storm to face.

After moving back home and getting our lives back on track, David and I decided that we wanted to fellowship at another church. We wanted to make a new start. We'd always talked about that anyway, but we knew when we returned to Boston, that's one of the things we both wanted to do. We ended up attending the Mt. Olive Kingdom Builders Church in Dorchester, where Bishop Robert C. Perry II is the pastor. I started my new job in February 2009. I was making very good money. David was the musical director at our new church, and we had two big celebrations for both of our daughters. We were in our own apartment. So, because of so much that happened in Texas, it took a big toll on us. Before we left Texas, David suggested that we seek a marriage counselor once we got back to Boston. I agreed—it was way too much to handle in Texas, and we felt that we would have a support system back at home. I'd always told David, and both of us agreed that no matter what happened to us, if we needed to talk to anybody, please make sure these people or person is neutral, and not biased, as that can become a big problem.

39. Speechless and Numb

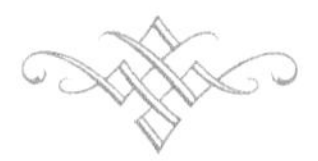

Some things, I wasn't prepared for. On the day of our wedding renewal anniversary June 2, 2009, David and I went out for dinner, to a place with a nice view near the water. For some strange reason, neither of us was hungry; we ended up ordering just dessert. We stayed until they were about to close, and then we got in the car. David started up the car, and the next thing I knew, he turned the engine back off. He said, "Saundra, we need to talk."

I said, "Okay, let's talk."

He hesitated over and over again, looked at me, and said, "I don't want to be married anymore—I never got a chance to experience life. I got married very young and it looks like your age may be a problem now. I miss not having my own biological child. I'm no longer in love…I have no feelings at all, and I want out of the marriage."

Well, that was a big pill for me to swallow all at one time! My mouth dropped to the floor of the car. It was a silent night for a moment, and I felt that what he had said was like a slap in the face. My heart dropped so low that I felt I couldn't pick myself back up—what harsh pain! I was never at a loss for words, but believe me, I did not know how to respond back to him. We both just sat in the car; maybe he was giving me time to digest it all. I was not sure. I believe he had thought all of this through ahead of time—it sounded like it had been going on for a while, even when I didn't see it coming. Yes, I had my say at the time—when it came to the child, I had plenty to say.

We'd talked about this before, even before we got married... my age! I guess at the time we met, it didn't matter? Not only that, I felt like my life got destroyed again, and the girls' lives as well. David and Kyva had a close bond. He came into her life when she was three years old, and now she was twenty-one—she considered him to be her father, even though she knew her biological father. This wasn't just about me; it was about our family being torn apart. We fought hard to be together in the very beginning, and now this was where it ended up.

After getting back home that night I felt no matter what I said, his mind was made up. I still loved David; therefore it was harder for me to deal with it. I always believed you took whatever measures were necessary to fight for your marriage, and after everything we'd gone through, I didn't think that I would have that kind of strength to hold on any more. Meanwhile, here came Valentine's Day, then our actual wedding anniversary, March 14th 2009. That was rough—David showed no emotional feelings at all. He was cold. David and I went to the same church, we were still living together, going on like it was nothing, but there was a lot of tension between the two of us. I was still working, up until later when I got injured on the job, when I went on workman's comp in December 2010. The money wasn't the same, but there was nothing I could do until I had surgery; perhaps I could find another job after healing. I did not let it faze me a bit that I was out of work—being on that job, working retail, I worked my behind off, sun-up to sun-down—it was very stressful. Then I would still come home, cook, clean, and do my mother and wife duties—that's just the way I am.

David and I sat down again and had a long talk. In the past, whenever there was tension in our marriage, or family, I got used to the fact that David and I would always pray, or pray with the family, but that fell off—he didn't want to pray together anymore, nor did he want to seek a marriage counselor. His mind was truly made up. I would always be the one to initiate to talk it out—whether he stayed or left, there had to be some kind of communication going. But prior to that, he stopped the communication—he stopped talking to me. He'd become very cold, and his whole attitude changed toward me. I would still prepare dinner for him, but he would come home and not eat. Things just got worse. I tried my best to stay out of his way—at night when we went to sleep; he would turn his face toward the wall and his back facing me, with a pillow between us, as if I were not there. I would turn over and start crying within myself, asking why this was happening to me again. I was tired of men ripping my heart out, and this time around was very hard, because we'd been married fifteen years, and together eighteen years—do you throw away all of that history together, like it didn't mean a thing?

I knew I had to be on my own again, but I had so much weighing on my mind. My parents lived upstairs on the second floor, but I chose not to get them involved; even though they could see what was going on, they never said a word to David about our marriage, but would always speak kindly to him. They didn't make him feel uncomfortable while he was still living with me. Kyva felt the tension, and I knew, at that point, that David and I needed to talk about arrangements. I knew that he was planning on leaving, and I had already accepted that in my heart. David would never be the one to take that step and figure things out together first—he would put it off, until I'd had enough.

One morning we took the whole day and sat down and talked. By that time, I'd already calmed down within myself, figuring that whatever happens, happens. David said his feelings were dead, he felt nothing. I swallowed that up again; I already knew he was leaving, so I just needed to know when. He mentioned so calmly that he wanted a divorce. I said, "Okay—if that's what you want, go ahead; I'll sign the papers. I'm not fighting anymore. I don't have that kind of strength anymore."

I told him he was being disrespectful to me; he leaves early in the morning, and come home late at night, maybe to avoid facing me. When he came home he was on the computer all night, sometimes till the next morning. He could never talk on the phone in front of me; he would always answer it, and then go out of the room, and the text messaging would beep all night long. Sometimes he would get up, leave the room, then check his messages—he would never leave his phone hanging around. There were times we would get up the next morning and pass each other without saying anything. I felt like he didn't even want me to say anything to him—not a word. It killed him to look at me, and it was getting worse. That was a wake-up call for me. I started to become very bitter towards him, because he kept going on in silence; therefore I knew it was time for him to go. He didn't want to be there, and he had stopped communicating with the girls.

Kyva was coming to me again. She said, "Mom, why are you putting up with this?"

I did not want to act out of flesh, so I approached him in the room one afternoon and said, "We need to talk. When are you leaving? I'm not going to let you keep disrespecting me anymore! You hurt me in so many ways, and I pray to God that you didn't cheat on me."

He said I threatened him, and he got scared; he said that I was going to kill him. I never had and never will use those words to him. I believe he started telling people he was close to about what he thought I said (it wasn't true). I still loved him.

Yes, I was angry and hurt, and I felt betrayed—he was my best friend, my husband, the love of my life—I thought till death do us part. I confided in him about everything in my life. But there is nothing new under the sun, and today I'm not sure if he was really listening all that time. Ladies, after all that, was I supposed to be nice about it? You tell me. Or was I supposed to be relaxed? I spoke to him very harshly, yes, to let him know enough is enough, I'd had it with his behavior, and we were not going to continue living this way—it wasn't good for either of us or for the girls. I told him, "If you want out, then do what you have to do."

I couldn't take him living there anymore like we were roommates—or maybe zombies is the word to use…it was so bad; way too much stress in the house, but he was just taking his time making his transition. I feel that once he made up his mind that he wanted out of the marriage, and made it known to me, he should have had his act together to leave at that point; he shouldn't have been trying to buy time for his personal convenience. It wasn't my job to feel sorry for him. Finally he sat me down and told me he had found a place to live. We both calmed down. Sometimes you just get tired of fighting over and over again. I think I just lost my strength altogether—meanwhile I was still doing the wifely duties. To those who say I was mean and hateful, and self-centered, and controlling—yeah, right! You are so wrong—and it looks to me like he's still alive, right? My name really went down the drain to some. He did leave for two days. I started to feel sorry for him; I knew that he wasn't ready to step out on his own yet, but he wanted out of the marriage so badly.

He did come by the house—he looked pitiful. We both started losing weight, so that wasn't anything new. He asked if he could take a shower. I said, "Go ahead—you made this choice, not me. If you need to shower or eat, I'm not going to turn you down…that's the kind of person I am. I would never turn a person away who needs help." So he ended up staying. Even though we were living like roommates, I managed to pull myself together, and not let my emotions get in the way.

40. God Will Bring You Through

I asked him to sleep on the sofa. I was still bitter, angry, confused, and disgusted inside, because he took our marriage this far without seeking help together—he didn't want anyone to help the marriage. You don't run to people who are against the marriage—that would be the wrong choice. I just could not believe this happened again—but worse, why did I allow myself to put up with this so long? David was still faithful in paying the bills; I will never take that away from him. I was a little scared with him moving out, because I didn't make the same amount of income, being on workman's comp. I said, "God brought me through everything else I went through in my life—why can't He bring me through this storm?" I took my wedding vows very seriously; now when the minister asks, "For better and for worse, richer and poorer," I know exactly what he is asking me—you have to really mean that, not just mumble words out your mouth. **Ladies, if a man doesn't understand his prize, someone else will!**

David did not give our marriage a chance to grow. Other people were always involved. When we relocated back from Houston, Texas, we did not seek wise counsel together, and we weren't praying. I had nothing to fight with, but I had to put it in God's hands, because what God joined together let no man put asunder! That's what the word of God says. David left the day after his birthday. I was trying to go away that weekend, and let him come pack up without me being there. I still was shocked, because David and I were never away from each other for Eighteen years, and we were married for fourteen years together—that's a long time together. My girlfriend Poma called me up and said, "Let me take you away for a day."

I said, "That will work—David is moving out, and I don't want to be around."

That Sunday morning she was on her way to pick me up, and I was excited to go, just get away for a minute at least. She called me up and said she was stuck in traffic—there was a big accident on the way. I said, "Oh no—well, just call me when you get closer." It was getting late. I waited on my first-floor porch. It was David's birthday, and Kyva and I had left him a gift on the table. It was weird not spending his birthday with him—we did all celebrations together. While I was waiting on the porch, a car pulled up—there was David and Bishop Perry.

I said, "Oh, no! She's not here yet!"

David got out the car, spoke, and went inside the house to change. I believe they were going out for David's birthday. Bishop Perry asked if I wanted to go, and said his wife was coming too! I said, "No thanks—I'm waiting for my ride to go away out of town. I'm all set."

He said, "You sure? You can come if you want to."

I know he felt bad, but it wasn't his fault. It was fine with me; I was getting away for a minute. I said to myself, *Dagg, girl! You are a strong woman!* But actually, it was nobody but God.

41. Getting Free

I finally got myself together, and not allowing my emotions to show, I was looking forward to going out of town. I needed to find somewhere very quiet and stress-free. No one knew where I was going except my parents and my two daughters. Well, I did just that—my girlfriend Poma and I had a wonderful time driving; that's where it all began. We got to our to destination, we pulled up, that's when I said, "The buck stops right here." I cleared my mind, and then I started letting go every bit of pain, hurt, bitterness, and anger—I just wanted to be at peace at least for a moment. It's like I wanted to cleanse myself out. We had a natural and spiritual experience. It was like hearing water running down the waterfalls—we played music, went out to eat, watched movies, prayed together, and encouraged one another. We laughed and laughed we even cried. I will never ever forget Poma. Yes, it was her—she allowed God to use her to help someone who was broken **inside and out**, and that one was me. You see, God always has a ram in the bush. I have such faith in God, and in order to find that peace within, you have to have a higher power to go to and speak to Him like I'm speaking to you. You can't fight a big battle by yourself—it's not your battle, anyway—and God will fight your battle if you let Him.

There were so many odds against our marriage; a lot of people just did not want us to be together, before, during, and after. We did end up going our separate ways—David pursued the divorce—but today I can say we try to respect each other we divorced January 2013. He still would do almost anything for me and the girls financially—that has never been a problem. I made up in my mind to this day, to give it all to God, and let Him be God. That's what I should have done from the start. God will never leave you nor forsake you. My girlfriend gave me a CD of a song titled, "Let Go and Let God." I played that song when I got home over and over again, until it registered in my sleep. When I woke up the next morning, finally I found a resting place within. You have to be around positive people. I would always make sure that my surrounding is good soil. I said to myself, *You really need to get it together, girl, and not allow yourself to fall apart anymore!*

I thought about all the **doors that I have gone through**, and realized it was time to say goodbye to those old problems, shake the dust off my feet, and say enough is enough. Everything I went through was another lesson to be learned, to help and teach other women what I did to keep my sanity. Not only that—I have to teach this to my daughters. I pray that they've learned from me. If you can prevent something from happening by learning from another person's lessons, then you should try not to go through those same doors. I pray that they don't have to go through what I've gone through. One thing I know is that they are strong.

42. Finishing What I Started

God has a plan for all of our lives, and it's up to us to seek our purpose in order to get to our destination. I feel that with everything I went through, one day someone will read my story, and be encouraged. Not only that, but they will say: "After all she's gone through, she's still standing strong, and at the end, she did not give up." Yes, it has to get better; God would put no more on you than you can bear.

Today I can truly say I'm free, and back to doing the things I got away from. I'm singing again, which I had put that on hold for a couple of years—if you don't exercise your talents and gifts, you will get rusty! I started back creating my fashions, hats, and other designs. I said, "The best thing for me to do is to finish what I started." I knew I should finish writing my book, which I started years ago. I had a lot of good things going on; when you allow things to sidetrack you, sometimes bouncing back is not so easy to do. Just know that the longer you wait, the harder it can be. I decided to shake whatever was holding me down. I got my prayer life back—that's first. I would call strong people that I knew to pray with me. They truly had my back, and would help pull me through it.

43. Why Not Forgive and Be Free?

Yes, I ask **God** to forgive me for not putting him first. I had to forgive myself for allowing these evil things to take control of my mind. I just started going down the line, and forgave the people that caused me so much pain and grief in my life. I rehearsed asking God to forgive me every day, until it just came naturally. I tried to keep thinking positive, and to keep all negative thinking out. Ladies, you have to love yourself first, before anyone can love you back! Just say **"I am fearfully and wonderfully made!" (Psalms 139:14)** Look at yourself in the mirror, and repeat that, saying it over and over again. Don't allow anyone to bring your self-esteem down—you are worthy, and more than that, God loves you more! **Hey, men out there,** if you have to repeat this for yourself, go ahead—we've all been through something, and are still going through the process. With all of that said, to this day I've never been without a roof over my head, I've never gone hungry, was able to buy groceries, my bills got paid on time, I pay my rent on time, and coming off my job, even though my income is not the same, I can truly say that I'm debt-free! I have a good paying job—look at God!

44. What Are My Daily Medicines?

When I go to bed every night, I pray. It does not have to be a long drawn-out prayer; just talk to God. I do that every morning and night. At night I would put on the gospel station, Daystar TV. They would always say the right things at the right time. I would listen to my gospel CD, and then I liked to really relax to smooth jazz, which is right up my alley. I love to listen to running water—I would keep my TV on Daystar at night around 4:00 a.m. That's when they play the worship reflection instrumental music—it's really soothing to your soul. It's instrumental only, with different sound effects, including waterfalls. That sends me off to sleep. When I get up in the morning, my voice teacher told me to start massaging my face and throat to relax, and then I start vocal training. It calms me down. My friend Robin told me to start pampering myself, just stay at a hotel by myself, treat myself, order something to eat, put a movie on—well, she took me away and did just that. I mean I spoiled my own self, and it was nice.

"Trust in the lord with all your heart and lean not on your own understanding." (Proverbs 3:5)

The other scripture that came to my mind is: "I can do all things through Christ which strengthened me." (Philippians 4:3)

I hope and pray that this story will help and touch whoever needs support. I have gone through many doors in my life. I still kept my peace of mind.

It's time to close these doors and never reopen them again!

45. Acknowledgments:

First to my lord and savior Jesus Christ, who's ahead of my life—I would be nothing without you. I wouldn't even be here, without your love, grace, and mercy. You get all the glory in this— because of you, I'm alive today, and I love you!

To my dear friend and sister, **Linda Taylor,** who made it happen for me—she helped me get a job! Yes, she is a friend, and a good listener—I will never forget our special friendship.

Thanks to my **Mother** (Mama Sue) and **Father** (Dat/Mr. Sonny) who never gave up on me, no matter what I was going through—love you both.

And to my precious gifted and talented daughters **Anisia** and **Kyva**—I will love you forever till death do us part!

My dear niece **Dominique** ,who keeps me with the fashions, shoes, and handbags— thanks, Nonick—don't stop!

Thank you **Willie Israel,** who also helped me get a job, during the bad economy in 2009; I am forever grateful.

To my Friend and Sister ***Angela Powell*** *who's close to my heart. A.K.A (high-up)*—you always keep it real with me. Our friendship will be forever—love you, girl!

Thank you to my brother **James W. Hills** for your support, and your wonderful job with my graphics and photos—you made it happen.

46. Inspiration & Blurb

Inspiration:

Saundra Mathis Copeland (affectionately known as Sandy) is a woman with great vision and insight. As she has multi-tasked through life, she has endured major setbacks that could have resulted in the abandonment of her dreams. Throughout her process, she persevered, never losing sight of her ultimate goal. The defeats could have been her elegy, but she transformed them into a melodic symphony; a triumphant celebration of victory. Her passion, resilience, and faith in God have brought her to this place. She is to be applauded for surviving the doors that attempted to stop her from succeeding. They may have delayed her, but they did not deter her. They became her launching pad that has catapulted her into her "set time." We shall watch, pray, and support her as she fulfills her destiny.

Wahseeola Evans

Blurb:

Saundra is an amazing, gifted woman of God who is sincere and is a blessing to share to all women who have gone through troubles, so they know they can stand firm, go though it all, and come out victorious.

Blessings,

(First) known as Danette Perry

Inspiration:

I, Sonji Neverson, met Saundra Copeland (Sandy Mathis) in high school. Back then I was a Muslim. I didn't know about gospel music, yet I knew that Sandy had an amazing gift as a singer. To my surprise, my parents welcomed a gospel singer into our home. Listening to Sandy sing "Amazing Grace" as my brother played the keyboard, my dream of singing grew.

One day, Saundra had the opportunity to hear me sing at @ home. I can remember Saundra encouraging me to sing my first solo. Saundra would say, "You can do it," and "I know you can." Today I'm excited to say that Jesus Christ is my Lord and Savior now, and currently I am singing gospel music.

Thank you, Saundra, for being an inspiration; for the prayers for myself and my family; and for being the first Christian example in my life.

47. Afterword

In *Many Doors,* Saundra has captured the essence of exploration through life by identifying the transitions we all have encountered. Although she has journeyed through different seasons in her life, the common thread is that God was there throughout all of them. He has kept her and has revealed His ultimate plan for her life. The fact that even through the tough lessons, God was with her is proof of His power and love reflected in her life. With every challenge God was with her. With every choice, God was with her and even in the chaos, God was with her. Now she has championed some of the areas of struggle in her life and yes God is still with her today. I believe this survivor's handbook will help many men and women to see that although they've gone through many doors in their lives, they too can be encouraged to know that as long as God is with them, they will survive!

Bishop Robert C. Perry, II
Pastor & Founder of Kingdom Builders' Worship Center

48. My Grand-Daughter Kayle Ny'El

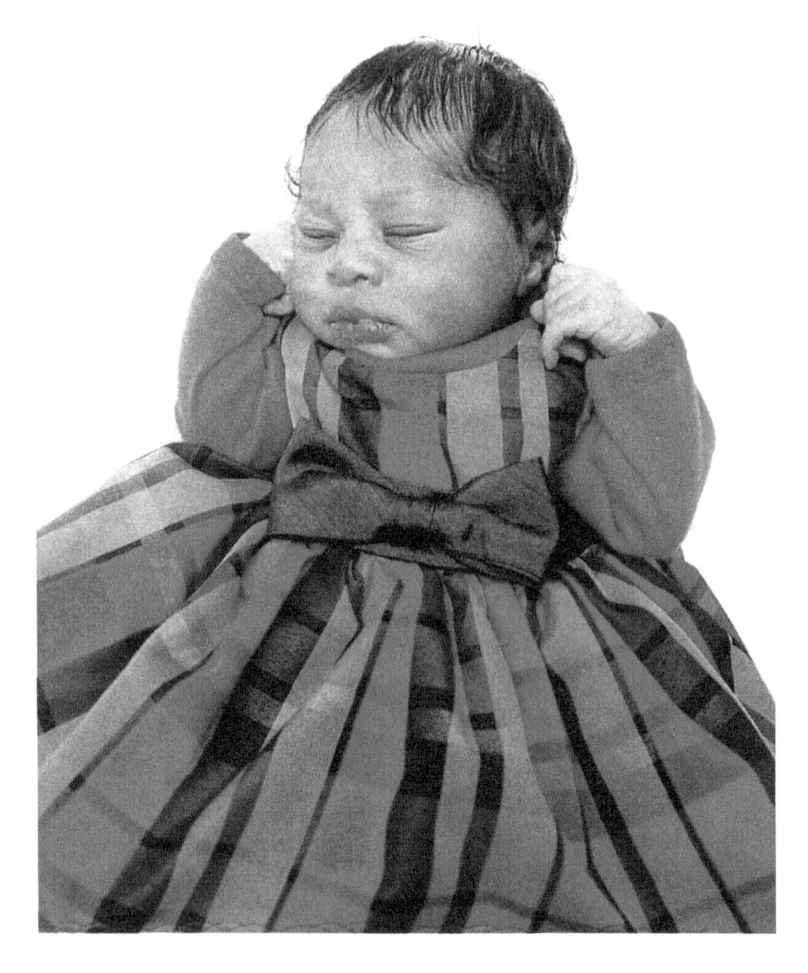

My Grand-Daughter

Here is another inspiration in my life, I would say the finishing touch of my Book my little Kaylee a.k.a (TIGER) I watched her mother give birth to her on 2/2/2013, and when she reached down, and pulled her all the way out, WOW! From that moment it did something to me, you see I have so much to live for, so here is my little Kaylee Ny'le

Love Mia

CPSIA information can be obtained at www.ICGtesting.com
Printed in the USA
BVOW11s2139060414

349710BV00003B/17/P